The War on Learning

The War on Learning

Gaining Ground in the Digital University

Elizabeth Losh

The MIT Press
Cambridge, Massachusetts
London, England

MIT Press books may be purchased at special quantity discounts for business or sales promotional use. For information, please email special_sales@mitpress.mit.edu.

This book was set in Stone by the MIT Press. Printed and bound in the United States of America.

Library of Congress Cataloging-in-Publication Data

Losh, Elizabeth M. (Elizabeth Mathews)
The war on learning : gaining ground in the digital university / by Elizabeth Losh.
pages cm
Includes bibliographical references and index.
ISBN 978-0-262-02738-0 (hardcover : alk. paper)
1. Education, Higher—Effect of technological innovations on. 2. Education, Higher—Computer network resources. 3. Teacher-student relationships. 4. Internet in education. 5. University extension. 6. Universities and colleges—Computer networks. 7. Educational technology. 8. Computer-assisted instruction. I. Title.

LB2395.7.L67 2014
378.1'7344678—dc23
2013038920

10 9 8 7 6 5 4 3 2 1

To my sister Janie. I fear she can still beat me in any word game.

Contents

Acknowledgments

This book describes the multiplicity of experiments taking place right now in higher education and the many ways that teachers and learners might be struggling to articulate their respective roles. Now that dramatically different power dynamics from those that constituted the modern university are being enacted in the halls of academia, when computational media on distributed networks are deployed, reaction rather than revolution often manifests itself. *The War on Learning* is a book that resists the acceptance of shortsighted commodity solutions from corporate vendors; instead, this book champions the making of hard choices about investment in new forms of digital labor and the adoption of new practices around digital literacy. I am grateful to my editor, Doug Sery, for helping me to see this ambitious project to completion.

This book benefited from having been in conversation with several critiques of the hidebound institutional culture of traditional universities. In particular, I am grateful to N. Katherine Hayles, who kindly shared a draft of her manuscript for *How We Think: Digital Media and Contemporary Technogenesis* very early in the writing process of this book. I also deeply valued our many conversations about pedagogy; her insights as a master teacher helped me at the practical level of writing syllabi and drafting assignments, but she also helped me at the theoretical level of refining my own claims about digital learning and instruction.

Lev Manovich provided invaluable assistance to me in revising the prospectus and rethinking the general approach of this project to make it speak more clearly to the existing literature of new media scholarship.

It's never easy to take risks as an instructor. I have been fortunate to benefit from the wisdom of many classroom innovators with whom I have presented workshops, given papers, and participated in roundtable dialogues about pedagogy that have refined my interpretative approach. The following early adopters shaped many of the ideas in this book: Jonathan

Alexander, Cheryl Ball, Kathi Inman Berens, Jamie Skye Bianco, Tom Boellstorff, Marc Bousquet, Benjamin Bratton, Jim Brown, Craig Dietrich, Jeremy Douglass, Hasan Elahi, Jason Farman, Kathleen Fitzpatrick, Armando Fox, Matthew Gold, Diane Harley, Katherine D. Harris, Ed Keller, Scott Kleinman, Scott Klemmer, Steven Krause, Virginia Kuhn, Antoinette LaFarge, Celine Latulipe, Erik Loyer, Mark Marino, Micki McGee, Dave Parry, Miriam Posner, Todd Presner, Stephen Ramsay, Alex Reid, Jana Remy, David Rieder, Daniel M. Russell, Mark Sample, Jentery Sayers, Trebor Scholz, Beth Simon, Karl Stolley, Doug Thomas, Annette Vee, Jacqueline Wernimont, Holly Willis, and Melanie Yergeau. I am particularly grateful to Jennifer Cool, who has probably read more pages of the manuscript in draft form than almost anyone else; she was a trusted audience for this project as a work in progress over many years. Although, to my knowledge, we have not appeared on any pedagogy panels together, Tracy Fullerton and Bill Griswold shared many documents from their projects, and our conversations about instructional technology are ongoing. Kim Yasuda and Marko Peljhan of the University of California Institute for Research in the Arts were also important drivers for pedagogical innovation in the institutional context of the UC system, and I always benefitted from our discussions about rethinking classroom dynamics and resources.

I also feel fortunate to have incredible colleagues at the Digital Media and Learning research initiative, especially David Theo Goldberg, Mimi Ito, Philipp Schmidt, and Nishant Shah. Although I no longer have an academic appointment at UC Irvine, I continue to be connected to the UCI intellectual network and the innovative scholarship being done there around digital learning, as it is broadly imagined to include many kinds of transactions, modes of collaboration, and instances of the cocreation of knowledge. I still consider Bill Maurer, Paul Dourish, Geof Bowker, and Garnet Hertz to be my colleagues, and I want to acknowledge their important contributions to my work, just as they have acknowledged feminist theorists in their own. I met many of the colleagues cited in this book thanks to Catherine Liu of the Humanities Collective at UCI; I am grateful for her friendship as well. Finally, among the UCI contingent, I was fortunate to share many conversations about digital pedagogy with Ava Arndt and her partner Sue Gautsch.

I owe a special debt of gratitude to the dozens of collaborators in the Dialogues on Feminism and Technology project, a distributed online collaborative curriculum that has involved a cohort of hundreds of scholar-teachers. Alexandra Juhasz first brought our classrooms into dialogue in 2008; she joined Anne Balsamo to start the network in 2012. I taught my first FemTechNet course in 2013 with the fantastic Lisa Cartwright, and I probably learned more about teaching in those ten weeks of team teaching and interdisciplinary instruction than I had in the past ten years.

I wish to acknowledge all the students I have taught at UC San Diego and UC Irvine who have written, blogged, designed, recorded, and coded such inspirational work in my courses. Thanks for laughing at my jokes and for being so patient and good humored, even when being experimented upon.

Special thanks are owed to all of my colleagues at Sixth College—especially the staff, faculty, and teaching assistants in the Culture, Art, and Technology (CAT) program, where I served as director during the time of this writing. CAT Associate Directors Diane Forbes Berthoud, Alexandra Sartor, and Eliza Slavet at Sixth are all educational innovators and researchers and, along with Lynette Brossard, Ethel Lu, and Marissa Martinez, were thoughtful critics on the topic of the challenges of effecting institutional reforms. Among CAT faculty, Patrick Anderson, Micha Cárdenas, Teddy Cruz, Ricardo Dominguez, Brian Goldfarb, Tara Knight, Jessica Pressman, Emily Roxworthy, Michael Trigilio, and Wayne Yang have been particularly influential in my thinking about how to spur greater student engagement by rethinking their relationship to digital sources. PhD and MFA students teaching discussion sections in the CAT program—particularly Kiik Araki-Kawaguchi, Joe Bigham, Marisa Brandt, Catherine Condliffe, Sophia Echavarría, Lorraine Graham, Heidi Kayser, Angela Kim, Keith McCleary, Kelli Moore, Chuk Moran, James Perez, Ben Segal, Ash Smith, and Tara Zepel—have also shaped my attitudes about instructional technology. As a lecturing faculty member, their advice from the trenches was incredibly useful to my teaching practice.

The labor of creating the actual manuscript, typed on an Apple laptop in a space that was once a sewing nook, would not have been possible without my husband Mel Horan.

Finally, many thanks to my sons, Felix and Tycho, for putting up with a writer-teacher-researcher as a parent. You have certainly taught me much more about the relationship between technology and learning than I have taught you. Good rhetoric speaks from the heart, takes the higher ground, and always is generous. May what we wish for in our interactions with our nearest and dearest, as we all learn together, help us interact with others in an increasingly technologically mediated world.

Introduction

When I tried to move into my new office, with the mission of teaching new media literacies, there was a sizable obstacle blocking my path. It was a printing press. It obstructed much of the hallway with its imposing mechanism, which consisted of a large wheel with six sinuous spokes and a variety of interlocking parts that included gears, levers, and rollers. There was a tray awkwardly stationed below this apparatus to contain any mess should the device ever be put into operation again. Yet it had been frozen in time since the printing of a celebratory announcement years before. The blocks of type had been left in the machine, and the ink had congealed around them. In its state of disuse, the machine's once-fluid assembly of articulated parts had become hardened and seemed to be incapable of ever again generating new texts.

The printing press had been installed in our offices as a reminder that the Internet was not the first information technology and may not be the last, and that moveable type had transformed society and given past populations the feeling that they were part of a time of dramatic change driven by technology. It was also hoped that this magical printing machine would unleash the creativity of the college's young do-it-yourselfers, who could then create their own scandal sheets to challenge the authorities or make broadsides to beautify the campus with their handicraft.

The printing press was a legacy, an inheritance. As the new caretaker, I soon discovered that it would be difficult to move the machine—much less to dispose of it. Heavy objects present a challenge to campus employees prohibited from undertaking duties outside of their job descriptions, particularly those tasks that could cause injury, and collective bargaining agreements guaranteed the official movers wages high enough to be out of the price range of my program budget. Although a museum of printing eventually took the press into its collection, there was a long period of time during which it was uncertain whether we would ever find another space in which it could rest.

Figure 0.1
The UCSD Sixth College printing press. Courtesy of Susie Bielak.

Meanwhile, those who wanted to protect the printing press from the perceived callousness of being discarded argued that we should serve as a site of sanctuary for this endangered device. They pleaded the case for providing refuge; they argued that the printing press was not, in fact, obsolete. They advocated for respect for the long history of information technologies, for showcasing new forms of study around comparative textual media, and for abiding by the ethical imperative to facilitate critical making in the academy. Their arguments have, since that time of exigency and forced disposal, become important to my thinking about what the humanities could mean for contemporary life in the digital age. But I don't regret overruling the machine's would-be saviors, because their arguments did not solve the problem of my having to navigate around the discarded machine and get into my office without incident.

Digital technologies are frequently imagined as being the very opposite of that printing press: new, cheap, light, compact, invisible, and labor-saving. Yet legacies such as my hallway printing press should be a reminder that information technologies function thanks to platforms that rely on the material world's physical constraints and affordances and operate in the context of users' aspirations, desires, and fears regarding the regimes of labor and property.

In other words, rather than contrast virtual reality with a reality composed of matter, it is important to keep in mind that digital signals are transmitted on machines made up of parts; cyberspace exists because of plastic, metal, rare minerals, and other substances that facilitate electronic communication. Earlier schemes to create a "University of the Air" out of seemingly ethereal radio or television signals still required investment in hardware for broadcast stations and transmitters,[1] and today's supposedly wireless campuses similarly depend on the stuff that makes up giant server farms and data centers.[2] The magical technologies that expand the global reach of the American research university still rely on huge transoceanic cables and the goodwill of station operators, fisherman, park rangers, and other custodians of the massive equipment assemblage that supports the Internet.[3] Satellite dishes that point toward the southern sky can shape our social relations in the northern hemisphere,[4] and the assembly of mobile devices in Chinese factories can influence digital practices in the United States.[5] Cracked screens or sounds of malaise may remind us of the mechanical and materially contingent nature of our devices. Rather than regarding bits and atoms as separate categories,[6] we should remember that magnetic-force microscopy reveals the physical presence of bits on the surface of a hard drive.[7]

At the theoretical level, taking a serious interest in material culture requires new ways of thinking about computational media and, consequently, new approaches to imagining digital learning as well. Three specific critical frameworks are important for the arguments made in this book: (1) object-oriented ontology, (2) media archeology, and (3) feminist theory, specifically feminist theories about boundary objects (chapter 5), infrastructure (chapter 7), and situated interactions (chapter 9). Theories about how technology is gendered might not seem obviously important to many stakeholders when approaching policy debates. After all, why should a seemingly neutral position on adopting new instructional tools be interrogated as a problematically masculine one? But feminist research on technoculture often uncovers hidden assumptions about gender, sexuality, race, class, age, and ability that allow us to better understand how we use, and are used by, machines.

I will argue that having at least some interest in *mess* as an area of inquiry is fundamental to understanding how technologies, people, resources, and networks work—and sometimes don't work—together. As computer scientist Paul Dourish and anthropologist Genevieve Bell write in their analysis of the cultural imaginaries of ubiquitous computing, "mess" reveals that "the practice of any technology in the world is never quite as simple, straightforward, or idealized as it is imagined to be" and that "technological realities are always contested."[8] Feminist scholars know that mess also has a politics of labor and property associated with it and recognize that new solutions sometimes create new problems. For example, when supposedly labor-saving devices were created for twentieth-century American households, women didn't find themselves with more leisure time. Instead, access to these new technologies created new expectations for higher standards of cleanliness, so that those primarily responsible for managing literal messes in the home became further and further behind in their duties as daily hygiene standards were recalibrated.[9]

If leaders in higher education aim to advance the telepresence and ubiquity of academic knowledge without suffering the consequences of their own hubris, they should be wary of supposedly rational strategic planning, hyperconfidence in the purity and abstraction of institutional logics, and the rhetorics of boosterism from the elect. Unfortunately, those responsible for dealing with the messy aspects of technology—aspects that could be perceived as negative obstructions to promoting a positive technological vision—can, much like those in certain traditional societies who handle excreta and death, become untouchables in the academy.

As a result, there is often too little thought invested in preparing for the eventual demise of a given instructional technology. After all, obsolete classroom devices, such as the filmstrip machine or the overhead projector, were once standard equipment for practitioners of the educational sciences, much as magic lanterns functioned in the classrooms and lecture halls of the nineteenth century.[10] Manuals on using filmstrip machines from as recently as the 1980s speak of potential gains in student attention, participation, and other behavioral objectives.[11] Guides to overhead projection from the same period praise the fact that the instructor can maintain eye contact with the audience and avoid darkening the room.[12] Both technologies became established after being deemed effective in military training, generated corporate profits with patented inventions, and then experienced a rapid decline in use with the rise of portable computational media that might, in turn, be eventually supplanted by even newer machines for learning.

John Law, a scholar well known for advocating the theoretical turn toward mess,[13] once observed that the overhead projector defined the terms for the social relationship between students and teachers because it was a technology capable of exercising a considerable amount of coercive power.

> I am standing on a stage. The students face me, behind seried ranks of desks, with paper and pens. They are writing notes. They can see me, and they can hear me. But they can also see the transparencies that I put in the overhead projector. So the projector, like the shape of the room, participates in the shaping of our interaction. It mediates our communication and it does this asymmetrically, amplifying what I say without giving students much of a chance to answer back. ... In another world it might, of course, be different. The students might storm the podium and take control of the overhead projector. Or they might, as they do if I lecture badly, simply ignore me. But they don't, and while they don't the projector participates in our social relations: it helps to define the lecturer-student relationship. It is a *part* of the social. It operates on them to influence the way in which they act.[14]

In Law's analysis, instructional technology shapes interaction, mediates communication, participates in social relations, and amplifies the message of the instructor.

I would argue that this amplification paradigm contributes to a mistake that advocates for instructional technology frequently make by imagining the situation of teaching as only a signal to be broadcast. Given the legacies that universities have inherited from early innovations in information science that emanated from places like Bell Laboratories and people like Claude Shannon, who wrote *The Mathematical Theory of Communication*,[15] it is not surprising that the signal theory of communication continues to be dominant in how educational media experiences are imagined. J. C. R. Licklider was correct in defining the computer as a communication device that also performs calculations,[16] but computers do much more than merely help transmit information through a noisy channel to a receiver.

Technologies don't just amplify messages; technologies also organize information, and thus they shape how we make meaning and classify the components in our world. What David Golumbia has called "the cultural logic of computation" can come into play, in which the calculations that operate in bureaucratic, centralized administrations can be deployed as mathematical simulations, although these calculated operations may be disguised by fictions, which he sees promulgated by Nicholas Negroponte and other technofuturists, about new forms of collaboration, the development of new perspectives, and the limitations of the traditional humanities. Golumbia writes that computation "can then be used, at sovereign discretion, as part of instruction, as a way of conditioning subjects to respond well to the computational model."[17]

Another mistake that many make is in believing that technology provides instruction more directly than other methods and that thinking about digital course materials from a media studies perspective doesn't contribute anything to the debate. Students are exposed to many forms of media, including textbooks and lecture slides, in traditional classrooms, but computational media, because of its programmable, procedural character, functions in a number of ways that distinguish it from other media. Computer-aided instruction may make obvious rhetorical appeals on the visual or verbal register, but the persuasive workings of its complex rule-based systems, as expressed by computer code, may be much more difficult to decipher. Information concerning what gets rewarded and how data is harvested may be made intentionally murky by online administrators trying to prevent students from gaming the system without actually learning anything, but the motivation for placing so much of the educational-content delivery mechanism inside an impenetrable black box of proprietary software is often driven by a quest for profit, bottom-line efficiency, and market share rather than the legitimate pursuit of genuine learning outcomes. Unfortunately, in gauging success, such enterprises often focus on the goals that are most easily measured in quantifiable terms, rather than focusing on the more meaningful results that are difficult for computers to calculate, such as those derived from sustained intellectual development and the application of theory to practice in real life.

Still another mistake made by digital learning initiatives is the tendency to assume that each project exists on a tabula rasa and that nothing carries over from print culture or from the existing institutional conventions of higher education—as though institutions can be transformed overnight to embrace a whole new assemblage of job holders, practices, beliefs, and objects. Anyone who has ever seen a ten-hour clock from the French Revolution should be persuaded that nothing basic changes so quickly. Even the QWERTY keyboard that students use on laptops and smart phones represents a link with the past: its organization of keys was intended to avoid jamming on nineteenth-century manual devices and to make the brand name "type writer" easy for salesmen to peck out on the top row.[18] However, if we were to embrace alternative keyboards customized for modern devices with supposedly more advantageously designed input interfaces, we probably wouldn't see much benefit in educational interactions, and expert users of the traditional keyboard would be the first ones to suffer.[19] Those contemplating a redesign of the Internet—to adapt it to the twenty-first century and the explosion of IP addresses, the rise of streaming media, and the need for strong encryption of critical assets—face similar obstacles to reinvention.[20]

It is difficult to ignore the forces of path dependence when considering the relationships between humans and machines,[21] particularly when one decision about instructional technology can influence many others.

Finally, there is the mistake of assuming that policy makers just need to get past the rhetoric of the digital university in order to make rational decisions based on pure facts about technology and user needs. As a rhetorician, I think this is an incredibly specious approach to debate and is only intended to silence opposing views. Deriding rhetoric is particularly ridiculous given the fact that the university is a rhetorical enterprise by its very nature. I happen to specialize in a growing subfield, usually called digital rhetoric, that informs many of the theories about online education developed in this book.[22]

Some colleagues of mine in the Digital Rhetoric Collaborative might define some terms differently than I do,[23] but we are all using the term "digital rhetoric" or "digital rhetorics" to make similar connections between rhetorical theory and rhetorical practice and to engage with policy and pedagogy actively, with an eye toward interesting failures and unlikely successes when it comes to teaching, learning, thinking, and deciding with technology. Such scholars are interested in both rhetoric *about* digital media and rhetoric that is conveyed *through* digital media. Rather than thinking about computational media or distributed networks for communication merely as tools, digital rhetoricians tend to think of them in terms of a series of affordances and constraints at work in public discourse. In other words, we are interested in platforms, interfaces, and code, and we are engaged with discourses about design and human–computer interaction. Like other kinds of rhetorical training, we value explicit instruction and believe in teaching students from a literature of specific subject matter, such as how to write code for computer programs or how to design effective graphical user interfaces. We also assume that all participants in the classroom are important rhetorical actors and that truth emerges in dialogue as knowledge is coconstructed.

In my own work, I often try to look away from popular hyperbole about digital influence that divides up into crisis-and-excellence binaries intended to capture public attention via appeals to ratings-driven mainstream media. Instead, I would rather look toward topics that might otherwise be overlooked in the dominant push to explain everything about computational media either with moral panics about insidious behaviors or vacuous cheerleading for any new technology regardless of its cultural value. This means that my work in digital rhetoric, when it has to be translated into a news story, is sometimes misunderstood.[24]

I have a particular interest in what game scholar Ian Bogost has called "procedural rhetoric," which may be quite distinct from "verbal rhetoric" or "visual rhetoric."[25] Bogost and others involved in the field of software studies are interested in the way that our use of computational media might not only influence but also engage us, as we navigate individual paths through games and other programmed experiences, in new forms of discourse that are different from traditional arguments. For example, such procedural rhetoric might serve to promote particular means of persuasion that are determined by our kinesthetic interactions with specific screens and input devices, or it might be concerned with how our limited experience of a computer's complex expressive functions are driven by specific algorithms in computer programming or specific platform choices in chips and operating systems.

In other words, in all of the examples of the "war on learning" that I present in this book—including online videos, educational video games, term-paper databases, and course management systems—I am interested in how the software and hardware is actually structured by Web designers, instructional technologists, and computer programmers. There is a rhetoric to online learning that requires knowledge of technical specifics as well as practical pedagogical application.

I've been involved in computer-aided instruction as a teacher since 1987, when I ran an afterschool computer lab at a delinquency prevention center in the community where I live today. We taught with the products of companies that have long since gone out of business, such as Brøderbund Software, or are barely surviving in the competitive educational software market, such as the Voyager Learning Company. From the time of my first teaching experiences in higher education as a graduate student in 1990, I have occupied a front-row seat for watching instructional technology take center stage in the academy. Sometimes the spectacle of scholarly technofuturism has been exhilarating; sometimes it has been deeply depressing. I have included a number of first-person accounts in this book. As a feminist and a rhetorician, I don't believe in erasing the scholar who bears witness from the scene, and I feel as a writer that narratives with descriptions, characters, and plots are more likely to engage public audiences subsidizing pedagogical experiments.

This book often tells the story of initiatives that fail because they treat education as a product rather than a process. Video games for the classroom, clickers for the lecture hall, laptops for the developing world, and consumer goods for teaching programming languages all let students down because they promote values of consumerism and consumption rather than other ideologies—such as intellectual development and scholarly participation—that don't fit with market models.

This book also continues themes that were developed in my first book, *Virtualpolitik: An Electronic History of Government Media-making in a Time of War, Scandal, Disaster, Miscommunication, and Mistakes*, which was about the government as a digital media-maker. In that book I argued that there is an inherent conflict between content creation and institutional functions of regulation—one that traditional public organizations representing knowledge, power, and history cannot ignore when they attempt to survive and remain relevant in the Internet era. I detailed many cases in which digital files reached unintended audiences and were used for unanticipated purposes, much to the chagrin of politicians and bureaucrats who wanted to maintain message control. Although that book is about government agencies and this one is about universities, a similar institutional desire for turning the Internet into a public-relations apparatus and avoiding coping with the inevitable failures that result from that policy is shown in operation in both books.

Debates about new media and learning have long rhetorical histories that go all the way back to Plato and Aristotle, who argued about the introduction of technologies like writing, the influence of media experiences like tragic drama, and the ways in which media technologies impact educational efforts either positively or negatively; given this history, I am not sure that the present moment can be easily extricated from its larger cultural conversation in ways that would allow instructional output to be measured objectively in simple numerical matrices and reductive schemes. Ultimately, all the arguments about learning and new media presented in this book have to do with meaning, cultural significance, and how the terms of a given argument are defined; inevitably, rhetoric has to play a major role. As we all search for a way to gain ground in the digital university, the debate over new media and learning is a case in which both the rhetoric about learning and the rhetoric of learning matter.

The first chapter of this book, "What They Learn in College," focuses on how cheating, modding, and hacking are powerful forms of informal learning even though they are fundamentally at odds with academic norms regarding honesty that are supposed to characterize university life. I argue that formal institutions of codified pedagogy, represented by universities, and informal learning among people who are largely self-taught appear to be increasingly in conflict; access to and use of computational media seems to exacerbate this tension. This situation is particularly unfortunate in the era of networked social computing, when one would hope that academic and popular forms of instruction would be converging. However, subversive forms of youthful behavior are instead often exoticized and rationalized as representative of a so-called digital generation that allegedly learns in a completely different way from students in earlier eras thanks to new, networked digital technologies that defeat traditional reading and conventional types of attention. A number of stereotypes about today's practically alien students also feed into utopian and dystopian narratives about a radical break from the past. I argue that online videos with how-to advice for students cheating on examinations and those that feature angry professors destroying the technologies of disobedient students are symptomatic of the supposed generational conflict putting the university into crisis. I also claim that clichés about generational divides are promoted by instructional technology companies that hope to profit from concerned educators by promising command and control of today's students.

The second chapter, "The War on Learning," examines how the rhetoric of contemporary debate about higher education has been framed in extremely binary terms in which there can only be absolute success or absolute failure, the eventual triumph of either the young or the old, and a sustainable future shaped by the mutually exclusive demands of either print culture or digital culture. I've seen lines being drawn in this supposed war on learning on my own campus, and I've seen how both sides seem to be defending asymmetries of power relations. I ask basic questions about recent moral panics concerning the young and also critique the utopian claims of self-described autodidacts who advocate for do-it-yourself (DIY) learning. In understanding the current "distraction panic" about multitasking students who resist the disciplining influence of educational institutions, I observe that lessons can be learned from even the most egregious moments of pedagogical failure. In unpacking university failure stories, including a memorable failure from my own lecture hall, I maintain that we must get outside the dichotomy of crisis and excellence to foster more reflection about how it is possible to have educational reform without educational revolution. After all, risk taking and managing controlled failure are required for learning because no student has perfect acquisition or retention of knowledge; if they did, any curiosity or engagement brought to the classroom would be extinguished.

My exploration of how failures produce successes and how successes may cover over failure continues in the third chapter, "On Camera: The Baked Professor Makes His Debut," by focusing on two case studies that I claim mirror each other: Howard "John" Hall and Randy Pausch. Both Hall and Pausch became Internet celebrities, with each starring in his own viral lecture video, but in one case we have a professor who embarrasses his institution by showing up for class apparently under the influence, and, in the other, we have a professor struggling with terminal cancer who is celebrated by his university for his laudatory public speaking performance and for the universal life lessons that his lecture seems to contain. However, I claim that further explication reveals a more complex set of rhetorical positions in these viral online lectures. Hall, in his disinhibited state, could be seen as defending his institution in the long term and providing a healthy challenge to it in his questioning of the university's corporatization and its embrace of an outmoded industrial model that no longer functions in the era of late capitalism. In contrast, careful reading of the text of Pausch's speech shows that he actually seems to argue that the rules of his institution can be broken to promote individual self-interest and that the corporatization of the university is not necessarily a bad thing. I then look at a range of online artifacts, created by various university stakeholders, that have come into public view and the anxieties that such user-generated content creates for universities, particularly when administrators don't know if unregulatable digital files will inspire awe or ridicule. I argue that sometimes the content that seems the least academic—a professor's online tantrum or an adjunct's parody—reveals the most meaningful lessons about the marketplace of ideas, the dynamics of power, and the counterintuitive qualities of knowledge-making in higher education. Although online education often celebrates the idea of the professorless universities of the future, I argue that professors are still quite visible online, and that they potentially serve as important rhetorical models for the conduct of students as well.

The fourth chapter, "From Reality TV to the Research University," examines a particular form of instructional technology—podcasting—and interrogates the rhetoric around the idea that university instruction in the form of modular units could be part of a typical student's playlist. By discussing the podcast as a genre, this chapter also introduces the distance-learning industry in higher education and explicates its hopes that mass-produced, video-recorded lessons can replace traditional live lectures, thereby serving larger numbers of students more conveniently and at lower cost. However, I argue that the public's expectations and our collective understanding of the appropriate reception of such modular audio and video content is often shaped by consumer patterns and specific norms of social behavior defined by the use of computational media. For example, the iPod and, later, the iPhone have influenced how we think about playlists, and expectations about online video have already largely been set by popular Internet hosting services like YouTube. In other words, when universities enter the realm of Apple, Google, and other corporate Internet portals, they risk being subject to the conventions of these sites, which includes everything from overt limitations on fair use, academic freedom, and student privacy to more insidious pressures to produce content designed for popularity and diversion. However, my criticism of online learning efforts in higher education, which draws heavily on comparisons with viral videos and reality TV, comes not from someone who hates popular culture, but from someone with a healthy respect for its appeals to broader audiences and who might even be considered a fan in some cases. To develop the argument, this chapter provides an extended explication of Tim's Podcast, which was produced in connection with the second season of the reality TV show *Project Runway* by former Parsons dean and department chair Tim Gunn. Chapter 4 then expands its analysis of new genres for online public speaking to look at techniques for self-presentation by university researchers in TED talks and how science is this presented to the general public as spectacle and narrative.

The fifth chapter, "The Rhetoric of the Open Courseware Movement" addresses the rhetoric of openness championed by some instructional technologists and how it often is not in practice at the very institutions considered to be its flagships. This chapter analyzes a range of position statements around open scholarly publishing, open access to course materials, open peer review, and open source technologies in learning management systems. It asks how these documents of the movement depict a particular model of knowledge acquisition and academic labor that may not be in actual practice anywhere, particularly when research shows that the next generation of college professors may actually be much more conservative about using digital technologies than boosters had predicted.

The sixth chapter, "Honor Coding: Plagiarism Software and Educational Opportunism," analyzes the adoption of plagiarism-detection technologies in universities and how particular ideas about academic honesty and the purity of single-authored texts can be interrogated. It looks at the complicated mesh of academic labor, intellectual property, and software procedures that play out when universities license the Turnitin plagiarism-detection algorithm, and it also examines how and why open-source alternatives have failed in the academic marketplace. It applies both contemporary appropriation theory from studies of remix culture, including the Citation Project, and the analogies between written language and computer code made by scholars in Critical Code Studies to argue that a new kind of writing is emerging that doesn't necessarily follow earlier theoretical models of hypertext, even if it is technologically complex and deeply embedded with practices of citation, commentary, and intertextuality.

The seventh chapter, "Toy Problems: Education as Product," examines the mass distributions of handheld digital devices to first-year students on college campuses. Such devices seem to offer new ways for undergraduates to "read" the university and its interfaces and to be instructed in new modes of digital reading by campuses interested in promoting electronic rites of initiation that are markedly different from those of the codex book. With mobile technologies like the iPad, students are encouraged to read textbooks, read buildings, and even read each other in ways that seemingly counter traditionalist anxieties about postliterate culture and the distraction panic. This chapter looks back on a decade of large-scale distribution efforts aimed at giving mobile computing to first-year college students and maps the trajectory that such products move through—from being featured in photo ops for press releases to being jettisoned as e-waste or surplussed by office personnel to eBay. To begin to understand the relationship between ubiquitous computing and learning, I describe the history of product-oriented cyberutopianism in my own Culture, Art, and Technology program at the University of California, San Diego, including an ambitious gadget-distribution program in which seven hundred wireless-equipped HP Jornada handheld Pocket PCs were given free to all our college freshmen in 2002. Although those devices quickly became obsolete, the architects of this "explorientation" may have actually understood more about how ubiquitous computing changes educational institutions than those in charge of many subsequent initiatives at other campuses throughout the following decade—a decade during which iPods and iPads have been handed out to first-year students as a way of transforming their college experience—because those who distributed the Jornadas a decade earlier also anticipated how social surveillance could conceivably backfire.

In the penultimate chapter, "The Play's the Thing: Games and Virtual Worlds in Higher Education," I argue that too much faith has been placed in an oversimplified understanding of the relationship between procedural systems and learning by advocates for instructional technology; I also argue that these cyberutopians have ignored key issues about how virtual classrooms may be designed, how texts may be adapted from one medium into another, or how interacting with the rule systems that govern interactive stories might send mixed messages to players—messages that work against more straightforward forms of didacticism. I examine a number of virtual reality environments for online learning and the exploration of shifting educational roles. In this chapter I also look at a number of recent experiments in teaching and learning in the online virtual world Second Life, many of which are now defunct, to think about social interaction, distance learning, and membership economies more broadly. I examine new trends in DIY higher education in which 2D badges rather than 3D costume regalia serve as the marker of status, and I consider how reputation functions as a measure of value independent of many other learning goals. I also question the rush toward gamification more generally. Several of the case studies in this chapter involve Shakespeare games that are intended to teach analytical principles and love of the Bard to young people. The longest case study in this chapter involves the disastrous development process of the multiplayer Shakespeare game *Arden*, which was funded, with high hopes, by the MacArthur Foundation and yet was ultimately dismissed by the project's principle investigator as a failure that lacked the critical element of fun.

The concluding chapter, "Gaining Ground in the Digital University," provides six basic principles and a number of case studies of successful university-based digital media and learning initiatives. It also suggests a framework for how to approach pedagogical experimentation ethically, use resources wisely, and think about technology more broadly so that projects can flourish after the first giddy launch period or—if appropriate—be allowed to die gracefully. Lessons learned from educational technology, including failures, must be preserved and become the inheritance of others. I make this argument as one who appreciates the virtues of messy forms of hybrid education rather than distance learning initiatives that attempt to erase the local peculiarities and quirks of specific communities of like-minded collaborators.

Digital learning initiatives often come connected to existing cultural practices and material constraints, and it is important to engage everyone—from self-declared Luddites to early adopters—during these transitional and uncertain times in postsecondary education. Computational media exist in social histories of membership, reputation, and attention, and even if the economies of labor and capital seem to be changing, the wealth or poverty of institutions in managing resources still matter in conventional terms.

I hope that college presidents will read this book. Unfortunately *The War on Learning* might not be appealing to this particular audience of a few thousand people, because it is a scholarly polemic about small negotiations rather than grand visions. I am concerned with the granular complexities of concepts in computational media, such as big data, telepresence, and ubiquity, that sound exciting in coverage of higher education in the news media but usually work quite differently from the way they are described in sound bites and press releases as bold adventures in educational entrepreneurship. In his classic work on instructional technology in the analog age, *Teachers and Machines*, Larry Cuban described how educational institutions tend to be subject to fads and yet resistant to reforms,[26] and I would argue that the same dynamic exists today. This is a book that tries to cover recent trends in instructional technology in frank terms, and I hope that its intended audience can appreciate this pragmatism. It is written using educational experiences that range from community colleges to large state institutions, from liberal arts colleges to design schools, and from recent for-profits to the establishment Ivy League.

I know that college presidents have grand visions, but they also have to make decisions on a daily basis about where resources are invested—and when and how and why and for whom. College presidents—like students and faculty—are often averse to experimentation and risk, so perhaps this book can help them better comprehend the rhetoric functioning around these ventures into the unknown and to understand technologically enhanced learning as a complex user experience in an intricate and fragile media ecology of rapidly changing forms of communication, algorithmic organization, and cultural memory.

1 What They Learn in College

"GMA," said the woman answering the phone on the opposite coast. The acronym was unfamiliar, so it took me a moment to realize that I had the right number for *Good Morning America* and was calling the correct person back from the popular ABC television show.

Months before, I had written an item on my blog about the existence of online cheating videos, created by high school and college students, that demonstrate elaborate techniques designed to boost examination scores.[1] The videos show, step by step, how to create fake drink bottle labels in Photoshop to hide formulae,[2] how to stuff pen shafts with answer scrolls,[3] and how to write detailed foreign-language conjugations on stretched-out rubber bands.[4] Most of the videos had low production values and were shot in informal domestic settings, but at least one Japanese video borrowed the actual form of commercial distance learning, with distinct chapters on the subject of cheating; the lessons were elaborated with slick information graphics and computer-generated animation to enhance the instructional content.[5]

My story about these academic dishonesty videos on YouTube had been picked up by a blog for the *Chronicle of Higher Education*.[6] Then a reporter from the *Shreveport Times* interviewed me and ran an article about the subject,[7] which later appeared in the large-circulation paper the *Chicago Sun-Times*.[8] From there, the story reached researchers at *Good Morning America* who thought it had the right kind of national, light-news appeal for early morning mainstream audiences. Although I am an academic, *Good Morning America* apparently hoped that I might be an adequately colorful commentator for the show and asked me to speak to the phenomenon of this new kind of online dishonesty.

I thought the cheating videos were interesting because they demonstrated an argument that I had been developing over the course of the past decade. First, with regard to everyday practices and long-term goals, formal institutions of codified pedagogy, represented by universities, were increasingly in conflict with individuals who were informally self-taught. Second, access to and use of computational media frequently seemed to exacerbate this tension. I thought this situation was particularly unfortunate in the era of socially networked computing, when one would hope that academic and popular forms of instruction would be converging to work in concert, thereby supporting a life-long culture of inquiry, collective intelligence, and distributed research practices. Certainly, the students in the videos were sharing tips and performing online knowledge-networking activities that constituted of a form of real learning, even if such learning would be considered objectionable by their professors who would, understandably, regard the content of the videos as being fundamentally in violation of the scholarly social contract.

In these cheating videos, the students demonstrated that they understood the procedures of their own educational institutions and had learned a different approach to the sequences of operations involved in the minutiae of test-taking—an approach that might guarantee higher scoring success. They had achieved mastery of something that seemed relevant to success in the university, but it wasn't what their professors wanted them to have learned.

In addressing the hundreds of thousands who watch such videos, students aren't the only ones in the implied audience. These videos appeal to many nonacademic viewers who enjoy watching, from a remove, the hacking of obstreperous or powerful systems as demonstrated in videos about, for instance, fooling electronic voting booths, hacking vending machines, opening locked cars with tennis balls, or smuggling contraband goods through airport x-ray devices. These cheating videos also belonged to a broader category of YouTube videos for do-it-yourself (DIY) enthusiasts—those who liked to see step-by-step execution of a project from start to finish. YouTube videos about crafts, cooking, carpentry, decorating, computer programming, and installing consumer technologies all follow this same basic format, and popular magazines like *Make* have capitalized on this subculture of avid project-based participants. Although these cultural practices may seem like a relatively new trend, one could look at DIY culture as part of a longer tradition of exercises devoted to *imitatio*, or the art of copying master works, which have been central to instruction for centuries.

Unfortunately, by the time the production offices for *Good Morning America* called my university office months after my initial blog posting, the story of these students' acts of ingenuity had become transformed by the producers into lurid evidence for yet another Internet moral panic.

Somewhere between the predator panic—when the news seemed to be all about online sexual deviants preying on innocent young people[9]—and the bullying panic over the cruelty of the young themselves toward their peers,[10] there were a variety of miscellaneous panics being aired on national television with horrified glee. My conversation with the GMA staffer quickly became strained. The show obviously wanted a voice of moral outrage to defend tradition and condemn the lax values of the young from the elevated position of a stereotypically highbrow college professor. Several times I was prompted to say how reprehensible I must think the behavior of the video makers was. At the same time, the woman from ABC was unable to suppress her voyeuristic pleasure in having lined up two of the actual video-producing student perpetrators so they could be paraded before the public in their unrepentant condition.

Ultimately, the episode of *Good Morning America* aired on December 10, 2008, without me.[11] The show included a segment with the title "High-Tech Cheating" that featured *New York Times* columnist Randy Cohen of "The Ethicist." Cohen was interested in the ironic mixture of exhibitionism and anonymity that these YouTube content creators valued, and—like me—he recognized the how-to genre being imitated by these aspiring YouTube stars.

> I was astonished at how skillfully they mimicked the form of the how-to video. They all had scores; they had soundtracks; they have opening credits; they have closing credits. … What was more striking about this than it being a record of their cheating was it being their desire I think to be television producers or television stars.

For Cohen, the students were primarily actors rather than educators; their frame of reference wasn't subverting educational institutions, it was emulating media organizations. Indeed, since many of the videos began with a disclaimer that cheating tactics were being shown "for entertainment purposes only," Cohen had good reasons to view the videos from this perspective.

Yet a number of university researchers have argued that cheating and learning are intimately connected in the digital era and that subverting a complex system is often the best way to understand its rules. Mizuko Ito and her team of MacArthur Foundation–funded researchers have studied a continuum of teen behaviors around knowledge transfer—from trading gossip to "geeking out" over shared enthusiasms for highly specialized topics.[12] Ito's group has noted that these kinds of informal educational practices are closely related to the widespread adoption of social networks and video games. As Ito and her collaborators explain, "access to cheats and other secondary gaming texts was common among kids."[13]

Prior to the release of this report, Mia Consalvo had argued that cheating in video games is expected behavior among players and that cheaters perform important epistemological work by sharing information about easy solutions on message boards, forums, and other venues for collaborations.[14] Consalvo also builds on the work of literacy theorist James Paul Gee, who asserts that video game narratives often require transgression to gain knowledge and that, just as passive obedience rarely produces insight in real classrooms, testing boundaries by disobeying the instructions of authority figures can be the best way to learn.[15] Because procedural culture is ubiquitous, however, Ian Bogost has insisted that defying rules and confronting the persuasive powers of certain architectures of control only brings other kinds of rules into play, since we can never really get outside of ideology and act as truly free agents, even when supposedly gaming the system.[16]

Ironically, more traditional ideas about fair play might block key paths to upward mobility and success in certain high-tech careers. For example, Betsy DiSalvo and Amy Bruckman, who have studied Atlanta-area African-American teens involved in service learning projects with game companies, argue that the conflict between the students' own beliefs in straightforward behavior and the ideologies of hacker culture makes participation in the informal gateway activities for computer science less likely.[17] Thus, urban youth who believe in tests of physical prowess, basketball-court egalitarianism, and a certain paradigm of conventional black masculinity that is coded as no-nonsense or—as Fox Harrell says—"solid"[18] might be less likely to take part in forms of "geeking out" that involve subverting a given set of rules. Similarly, Tracy Fullerton has argued that teenagers from families unfamiliar with the norms of higher education may also be hobbled by their reluctance to "strategize" more opportunistically about college admissions.[19] Fullerton's game *Pathfinder* is intended to help such students learn to game the system by literally learning to play a game about how listing the right kinds of high-status courses and extracurricular activities will gain them social capital with colleges.

The kind of learning associated with cheating, modding, and hacking also raises provocative questions about the set expectations of everyday life for those in more privileged groups. For example, my own son, who was nine years old at the time that I began writing this book, was playing a sequel to *The Sims*, which is a video game where players build houses, spend money, and do a number of everyday adult activities in the pursuit of wealth and happiness. Some versions of the game also give players god-like powers to design an entire world in which points are accrued based on the sustainable homeostasis of their model of utopia and the interactions between competing interests that the player has constructed. *The Sims* is a computer simulation that is often praised by educators who promote digital media and learning because the game requires a lot of problem solving. Although created by the famed libertarian Will Wright, *The Sims* launched a franchise in which players can participate in the social engineering of everything from the acquisition of consumer goods in an individual household to building large-scale urban infrastructures that include power grids and transportation policies intended to support the health of complex, globalized economies.

My son's digital media and learning experience with *The Sims* included several hours spent carefully choosing furnishings for the house of his virtual family. He didn't splurge on the most lavish electronics or opulent furniture for his computer-generated dwelling; his aim was to combine his aesthetic sensibilities about what would constitute good taste in a home while accommodating all the various needs of his simulated family (who were named after the members of his real family and shared many of their characteristics). However, he was soon disappointed to realize that he didn't have enough money in the world of the simulation to purchase the house that he had worked so hard to prepare for occupancy. Quickly he demanded to be allowed online to see if there were any cheat codes to remedy the situation.

Even from his nine-year-old perspective, cheating in the simulation was a rational response to the constraints arbitrarily introduced into the game to create want and competition. There was no scarcity of material resources dictating the situation he found himself in. Given the nature of the means of production in bits and bytes, a well-appointed virtual house costs no more in statistically significant terms to the company that publishes the game, Electronic Arts, than an impoverished dwelling. And he had already paid for the game with a birthday gift card that represented real currency and had contributed several hours worth of his labor in creating his ideal decorative scheme. Besides, he argued that his aims were altruistic in that he had planned to gift me the house to show me all the pleasures that could be had in a *Sims* environment, as a kind of vacation home that could serve as a refuge from the real-life drudgery that he saw was associated with a real-life home and its domestic chores.[20] Although the game emulates some of the unfairness of the real-life market system that excludes many from home ownership by creating barriers, the game's barriers could be easily overcome, unlike many barriers to social advancement—those of class, education, race, ethnicity, and language—in the real world. As game scholars McKenzie Wark and Jane McGonigal have both observed, games are appealing because the rules seem more fair and more oriented toward rewarding effort and problem-solving than the rules of the much more dysfunctional real world governed by consumer capitalism (Wark's critique) and largely disconnected from concerns for the personal happiness of individuals (McGonigal's area of concern).

However, Gee would later argue in *The Anti-Education Era* that gamesmanship that enables universal access and personal privilege may actually be extremely counterproductive. Hacks that "make the game easier or advantage the player" can "undermine the game's design and even ruin the game by making it too easy."[21] Furthermore, "perfecting the human urge to optimize" can go too far and lead to fatal consequences on a planet where resources can be exhausted too quickly and weaknesses can be exploited too frequently. Furthermore, Gee warns that educational systems that focus on individual optimization create cultures of "impoverished humans" in which learners never "confront challenge and frustration," "acquire new styles of learning," or "face failure squarely."[22]

In understanding the GMA segment on "High-Tech Cheating," it is worth noting that Cohen's "The Ethicist" column often deals with the ambiguities of everyday situations in which fairness, access to information, reciprocity, and obligation to abide by common social codes come into play, so the justifications that my nine-year-old provided might be familiar to him—even if cheating in a *Sims*-style game is clearly more of a victimless crime than cheating on an examination. For example, a classic case addressed in Cohen's column involves whether or not it is ethical to move up to more expensive seats that are unoccupied at a sporting event. In that case, decision-making to maximize resources and individual happiness is at odds with the variable reward structures of ticket pricing. As the one talking head introducing moral ambiguity on the show, Cohen was drawing viewers' attention to more subtle aspects of the cheating videos than the obvious fact that they condone cheating on examinations. If Cohen were speaking as a rhetorician rather than as a newspaper columnist, he might ask: What perfectly legitimate rhetorics do such cheating videos promulgate, and what conventions about instruction do they promote?

The *Good Morning America* story about the cheating videos was introduced by anchor Chris Cuomo, who explicitly connected the story to the Internet's generally understood role in online teaching and learning.

All of you know that the Internet has revolutionized the way that our kids learn. Of course you do. But did you know that it has also revolutionized the way that they cheat? Listen to this: 40 percent of American students admit to cheating. And Internet plagiarism, that used to be the big worry, but now there is a bigger concern:

online instructional videos, kids literally teaching other kids tricks to help them cheat at school.

After this reminder about the revolutionary character of the Internet, correspondent Juju Chang then picked up the story and compared knowledge acquisition in online communities to an "echo chamber" of received opinion.

And you can chalk this up just when you think you've seen it all online. Well, you know, cheating, of course, is nothing new, but what's alarming is that it is being celebrated in the echo chamber of the Internet. That's when it's time for adults to take a second look.

Later, contributor Ann Pleshette Murray, who was billed as an expert on parenting, complained lamely about how the "YouTube craze" encourages "shameless" behavior such as "boasting." Both YouTube cheaters that the episode profiled, Kiki Kho and Nate Igor Smith, blamed the viewer. For example, Kiki argued that she never "promoted cheating" and that it would be "the viewer's fault" if her advice was heeded. Rather than focus on why such vulnerabilities exist in a teach-to-the-test educational system, the segment closes with banal moralizing about how such cheaters only cheat themselves.

What's striking about the ABC coverage is that it lacked any of the criticism of the educational status quo that became so central for a number of readers of the earlier *Chronicle of Higher Education* story—those who were asking as educators either (1) what's wrong with the higher education system that students can subvert conventional tests so easily, or (2) what's right with YouTube culture that encourages participation, creativity, institutional subversion, and satire.[23] For example, DrFunZ puts the blame squarely on his own professorial class:

Students somehow believe that if they have the equations in front of them, or a bunch of facts to cheat from they will benefit during an exam. NOW, where exactly did they get that idea? They got that idea by realizing that it sometimes works!! Why does it work? It works when we in the professorate write exams that test simple facts or the ability to just plug in numbers, we make it easy to cheat. But, when we make exams that are more challenging, ones that required students to apply equations/facts to completely novel solutions, then this method of cheating would go extinct.

Most of the physical science profs in our department allow the students to use texts, notes and internet when doing exams. The exams are brain-burners—all application to novel problems. Rarely will more than a few get an A on the exams. Our mindset is like this: Just knowing formulas and facts is comparable to tossing your legs out the bed in the morning. Yes, you are up and awake, but you haven't really done anything yet. The real work is yet to come.

In contrast, another person posting comments, Allison, identifies herself as "an arts and crafts person" and expresses her admiration for "the quality—good pictures/right level of detail—of this training video." Commentator Peter Naegele complains that this kind of story about YouTube could easily be absorbed into "'technology is making students lazy and stupid' propaganda" despite the fact that the lazy ones might be the teachers rather than the students.

Many of the educators argued that in the age of ubiquitous computing, depriving students of reference materials does little to test them on their ability to apply concepts in real-world situations. Several provided testimonials to the value of allowing them to bring their own study aids, such as note cards, to examinations. As Herb puts it,

> The creation of cheat notes is actually an aid to remembering stuff in the first place. You've had to read, organize, re-type, and have things easily at hand when needed. Once you've done all that, the likelihood is that you'll remember stuff without the notes anyway. Tell your students they can bring to the test all the information they can put onto an index card, and watch them learn more.

This attitude reflects current research on so-called distributed cognition and how external markers can help humans to problem solve by both making solutions clearer and freeing up working memory that would otherwise be tied up in reciting basic reminders. Many of those commenting on the article also argued that secrecy did little to promote learning, a philosophy shared by Benjamin Bratton, head of the Center for Design and Geopolitics, who actually hands out the full text of his final examination on the first day of class so that students know exactly what they will be tested on.

At least one person who posted a response to the *Chronicle* article complained that this shortsighted behavior on the part of both test-takers and test-givers could be traced to the emphasis on multiple-choice testing during the administration of George W. Bush after the passage of the No Child Left Behind law in Congress. At the time that the cheating videos were being debated in the *Chronicle of Higher Education*, academics focusing on research in higher education were also discussing the so-called Spellings Report from the then-Secretary of Education Margaret Spellings, who argued for more multiple-choice testing examinations in college to create more quantifiable measures of success.[24]

Pressures to proceduralize education are nothing new. I well remember the drudgery of being an elementary school student forced to work methodically through the color-coded Science Research Associates (SRA) reading cards and their multiple-choice tests. Such cards had been adopted by school districts eager for standardized instruction and measurable results. I recall how my teachers would praise the SRA system as a form of instructional technology that would supposedly prevent quick readers like me from becoming bored by group instruction with slower readers and how the colorful cards and pride of progress were supposed to keep me more motivated than would more amorphous human-delivered lessons without clear benchmarks. The Science Research Associates company was founded in 1938 and acquired by IBM in 1965; it now offers online tutorials as part of the McGraw Hill textbook-publishing empire.[25]

Access to specific types of computational media, now possible in the era of personal and mobile computing, certainly changes the student-teacher dynamic significantly, and the equation of interactivity with engagement is driving educational policy in new ways. In particular, an entire academic discipline devoted to assessing student engagement has blossomed in recent years, as university administrators struggle to recruit and retain students who are tempted to check out for more fulfilling life experiences online. Reports from the National Survey of Student Engagement,[26] the Community College Survey of Student Engagement,[27] and the High School Survey of Student Engagement[28] explicitly make the issue of student engagement a top priority. Well-intentioned administrators may even construct learning communities that are so good at engaging students that they might have little to do with engaging faculty; such communities function instead as self-contained bubbles devoted exclusively to the jurisdiction of student affairs and residential life. Skeptics might even argue that the student-engagement movement encourages the pursuit of the pleasure principle—instead of actual learning objectives that equip students for demanding professions like engineering, medicine, and law. Certainly, even if this movement exists to bring the dorm room and the classroom closer together, to focus exclusively on engagement risks ignoring the existence of potential conflicts between what students want and what faculty want.

Us versus Them

This book explores the assumption that digital media deeply divide students and teachers and that a once covert war between "us" and "them" has turned into an open battle between "our" technologies and "their" technologies. On one side, we—the faculty—seem to control course management systems, online quizzes, wireless clickers, Internet access to PowerPoint slides and podcasts, and plagiarism-detection software. On the student side, they are armed with smart phones, laptops, music players, digital cameras, and social network sites. They seem to be the masters of these ubiquitous computing and recording technologies that can serve as advanced weapons allowing either escape to virtual or social realities far away from the lecture hall or—should they choose to document and broadcast the foibles of their faculty—exposure of that lecture hall to the outside world.

Each side is not really fighting the other, I argue, because both appear to be conducting an incredibly destructive war on learning itself by emphasizing competition and conflict rather than cooperation. I see problems both with using technologies to command and control young people into submission and with the utopian claims of advocates for DIY education, or "unschooling," who embrace a libertarian politics of each-one-for-himself or herself pedagogy and who, in the interest of promoting totally autonomous learning in individual private homes, seek to defund public institutions devoted to traditional learning collectives. Effective educators should be noncombatants, I am claiming, neither champions of the reactionary past nor of the radical future. In making the argument for becoming a conscientious objector in this war on learning, I am focusing on the present moment.

Both sides in the war on learning are also promoting a particular causal argument about technology of which I am deeply suspicious. Both groups believe that the present rupture between student and professor is caused by the advent of a unique digital generation that is assumed to be quite technically proficient at navigating computational media without formal instruction and that is likely to prefer digital activities to the reading of print texts. I've been a public opponent of casting students too easily as "digital natives" for a number of reasons. Of course, anthropology and sociology already supply a host of arguments against assuming preconceived ideas about what it means to be a native when studying group behavior.[29] I am particularly suspicious of this type of language about so-called digital natives because it could naturalize cultural practices, further a colonial othering of the young, and oversimplify complicated questions about membership in a group.[30] Furthermore, as someone who has been involved with digital literacy (and now digital fluency) for most of my academic career, I have seen firsthand how many students have serious problems with writing computer programs and how difficult it can be to establish priorities among educators—particularly educators from different disciplines or research tracks—when diverse populations of learners need to be served.[31]

In many ways, as the product of a time when the BASIC computer language was widely taught in schools, it could be said that I am much more digitally literate than my own students. Today there are many competing computer languages tailored for the K–12 environment, and programs are launched and dropped depending on the momentary enthusiasm of individual teachers. Thanks to after-school or summer programs, some students from more affluent schools or "geeky" backgrounds may be exposed to creating code in Scratch, Alice, Scheme, LSL, or ActionScript, but at other schools without equipment for hands-on work or teachers with enthusiasm and expertise, digital literacy often means learning to make tedious PowerPoint presentations or create inane iMovies rather than ever learning anything about writing code and authoring programs that can execute commands successfully without bugs. Even though relatively accessible computer languages like Processing have large user bases that create huge libraries of code that can run robots, remix video and music, and create beautiful animations and data visualizations,[32] many schools are actually cut off from access to such adult communities of online learners because of parent hysteria about Internet predators, cyberbullies, and "time wasting."[33]

In the book *Born Digital*, Harvard Law School professors John Palfrey and Urs Gasser attempt to present a balanced approach to a wide spectrum of online learning venues. Overall, they make a positive effort to engage both educators and students in the conversation by including both costs and benefits of new technologies, addressing both young and old, and getting beyond hype about gadgetry and youth rebellion to focus on the digital dossiers being created on all citizens. Thus Palfrey and Gasser attempt to move the focus from broad generalizations about kids today to the more subtle ambiguities involved in topics such as privacy and consent. Despite my respect for Palfrey as a colleague in the field of digital media and learning, I think he might still rely on four potentially destructive cultural clichés about how "they" are different from "us": (1) "They all have access to networked digital technologies. And they all have the skills to use these technologies"; (2) "And they're connected to one another by a common culture"; (3) "They are joined by a set of common practices"; and (4) "Digital natives can learn to use software in a snap."[34] Although the rise of mobile computing on cell phones has enhanced the technological abilities of many youth from low-income homes who grew up without access to desktop computing,[35] the ability to tinker meaningfully becomes increasingly difficult with such black-boxed devices that rely on business models that perpetuate a push toward miniaturized components, planned obsolescence, proprietary code, and noninterchangeable parts. I might say to Palfrey that there are huge curricular costs to assuming knowledge transfer without obstacles, and that the term digital natives has been widely adopted by news outlets relatively uncritically,[36] even as researchers point to how poorly digital natives may score on tests of basic digital skills, many of which depend on an ability to read fine print rather than respond to showy technical wizardry.[37]

Much of the argument for the existence of a digital generation was actually made many years earlier by those who commented on the rise of so-called "television children" who had supposedly become dependent on an earlier kind of electronic screen. Such young people could no longer respond to traditional learning delivery systems, Marshall McLuhan insisted, and would be more likely to thrive if they were taught by modern closed-circuit televisions with specialized programming.

> Our entire educational system is reactionary, oriented to past values and past technologies, and will likely continue so until the old generation relinquishes power. The generation gap is actually a chasm, separating not two age groups but two vastly divergent cultures. I can understand the ferment in our schools, because our educational system is totally rearview mirror. It's a dying and outdated system founded on literate values and fragmented and classified data totally unsuited to the needs of the first television generation.[38]

McLuhan's assertions about how to address the needs of a new generation might have seemed highly logical in the context of this 1969 *Playboy* interview, since a body of work from a number of different researchers who were exploring both the dangers and possibilities of contemporary media had already accumulated. A 1962 government report listed a large bibliography of studies with titles such as "An Investigation of Closed-Circuit Television for Teaching University Courses," "The Potentialities of Closed-Circuit Television: Teaching in Colleges and Universities," and "Closed-Circuit Television as a Medium of Instruction at New York University."[39] Researchers examined the more open medium of broadcast television as well, as the presence of "An Experimental Study of College Education Using Broadcast Television" indicates. According to McLuhan and other enthusiasts, younger consumers needed "hot" media that engaged all the senses rather than "cool" media such as print that required pattern recognition and the supply of missing information by the reader.

Some scholars argue that the empirical evidence for the existence of a digital generation is notably absent and that this rhetoric may have a number of other practical consequences beyond the ones that I have outlined. For example, Siva Vaidhyanathan argues in a 2008 essay called "Generational Myth" for the *Chronicle of Higher Education* that "not all young people are tech-savvy." In his opening paragraph he notes the dystopian and utopian rhetoric that constitutes the terms of the debate:

> Consider all the pundits, professors, and pop critics who have wrung their hands over the inadequacies of the so-called digital generation of young people filling our colleges and jobs. Then consider those commentators who celebrate the creative brilliance of digitally adept youth. To them all, I want to ask: Whom are you talking about? There is no such thing as a "digital generation."[40]

Vaidhyanathan draws on his personal experiences as a faculty member to point out that each of his classes has "a handful of people with amazing skills and a large number who can't deal with computers at all," and quotes my own reservations about digital literacy in the article. He notes that libraries and traditional print culture are thriving (and struggling) among people of all ages. He also makes a broader point that "generations" tend not to function well as descriptive cultural categories to characterize historical periods of change.

In the case of technologies associated with computational media, I think it is quite interesting to contrast older early adopters who experimented with new technologies for teaching and creating media, sometimes decades before these practices become widespread, with younger members of a more unevenly prepared digital generation in need of mentorship. Ironically, many of those early adopters are exiting the classroom as they reach retirement age, so the trend may actually be toward computational ignorance rather than computational enlightenment as time goes on. I also argue that the digital generation functions more as an ideological construct than as a demographic label for a group of subjects who can be studied empirically. When I talk about the myth of the digital generation, which is perpetuated about a supposedly identifiable and technologically empowered digital generation, I mean to use the word "myth" in both senses of the word: as a refuted falsehood and an origin story. Recent research in the social sciences may indicate that there are significant deficits in computational literacy among younger computer users. But a "myth" is also profoundly true, because of its explanatory power.

Promoters of gadget-oriented instructional technology argue that students in the digital generation need to be connected to a technical device in order for there to be any hope of engagement with formal education. Consider the home page of "Engaging Technologies" as an example of how the current instructional technology movement speaks about "educating a new generation." In a page of links to what seem to be research findings about "Why Use Clicker Technology?," the company boasts about the benefits of their technology specifically in reaching the young:

> Each student in your class has a wireless handheld response pad, or clicker, with which they are able to answer questions. You, the teacher, pose a question verbally

or through the computer onto a projector or television screen, and the students respond with their handheld device. Clicker technology, very similar to the familiar feel of a remote control in their hands, is comfortable and fun for the Net Generation. They instinctively know how this "gadget" works, and it makes learning fun. After students respond, you can then *see immediately how students answer* and if they understand the material being taught. This allows you to remediate if needed or forge on ahead with new material if it is clear that the students are ready for it; clickers provide true data-driven instruction. The ability to pose verbal questions and receive immediate feedback from the eInstruction clicker will also allow you to totally change the dynamics of that usually tiresome lecture time. *The classroom suddenly becomes engaging and interactive,* and your instructional time becomes much more effective.[41] (Emphases in the original)

Notice not only how engagement and interactivity are praised and conflated, but also how the rhetoric of novelty in consumer electronics and of short attention spans also comes into play.

Despite being busy stage-managing increasingly complex PowerPoint presentations or elaborate clicker quizzes recommended by instructional technologists, professors notice student apathy and preoccupation, and their feelings do get hurt. Bring up the subject of student engagement and digital distraction at any faculty gathering and brace yourself for tales of woe involving laptops, cell phones, and other instruments of pedagogical subversion that seem to further distance instructors from their students. When the *New York Times* ran a series of articles on the theme "Your Brain on Computers," public response was particularly strong to a story called "Growing Up Digital, Wired for Distraction." Some of the teachers in the article had complained that administrators were pandering to shorter attention spans. "When rock 'n' roll came about, we didn't start using it in classrooms like we're doing with technology," a particularly forlorn Latin teacher lamented.[42] Unlike the predator panic or the cyberbullying panic of recent years, where the older generation could step in and protect the young from their own computers, the distraction panic posits that there is an unbridgeable generational divide that creates a potentially disastrous rupture in norms of the classroom and, inevitably, in the fabric of society as well. Because of the inability of authority figures to contain the dysfunction, distraction seems to be a problem with no foreseeable solution.

In contrast, some scholars have argued that this inability to focus might actually be preservative in an age of rapid-fire distributed stimuli. For example, Duke professor Cathy Davidson has defended multitasking as a way of protecting against the dangers of "attention blindness" in which sustained attention with a narrow focus might actually cause people to miss crucial information in the larger scene. In *Now You See It: How the Brain Science of Attention Will Transform the Way We Live, Work, and Learn,* Davidson asserts that we should learn the central lesson of a famous cognitive experiment in which research subjects told to count the passes of a basketball miss the fact that a person dressed in a gorilla suit enters and exits the center of the basketball court scene that they are intently watching. According to Davidson, "our digital age demands a different form of attention than we've needed before" now that our "primary information source is Google, where a search for information about 'attention' turns up 371 million entries, and there's no librarian in sight."[43]

Everyone Behaving Badly

There is a saying among people in the instructional technology movement: "distance learning begins in the second row."

Anyone who has observed a traditional, large lecture hall course in the past decade recognizes the truth of that statement and the serious lack of empathy or emotional connection that many students in the back row of the lecture halls of today feel toward their physically remote college instructors. From the cynical perspective of these highly disengaged students, professors might just as well be audio-animatronic robots in the higher education theme park in which they plan to spend the next few years in merriment and diversion with their friends. No wonder such students would band together to make cheating a rite of passage, the logic goes, when professors seem unimportant, and test scores are the only things that matter.

Professorial hostility toward students anywhere beyond the first row also seems magnified by access to technology. Type the words "angry professor" into a search engine and be prepared for videos of professors destroying digital devices or chiding their students for inattention. As the many remixes, parodies, and hoaxes generated around these angry professor videos demonstrate, trying to police student distraction in this way often only seems to "feed the trolls." What might be less obvious to the uninitiated is that many of these videos showing supposedly inflamed faculty performances online are actually hoaxes or spoofs. Some, such as "Teacher Breaks Student's Phone," now at over a million views, have obviously canned dialogue and are quickly called out as phony by viewers.[44] "'I Have a Strict Texting Policy' (Professor Destroys Cell Phone)" has earned almost seven hundred thousand hits, despite clues to its falsity such as a faculty member looking at the camera awkwardly and a history lesson that would be too simplistic for most high school classes.[45] Nonetheless, a campus newspaper took the "texting policy" video as genuine and ran a related story about class policies and disciplinary consequences that such a professor might face for destroying student property.[46]

Other videos may have more complicated reception histories if they have been passed on through online social networks as genuine evidence of faculty retaliation for exasperating student behavior. For example, Professor Kieran Mullen, a condensed matter theorist at the University of Oklahoma, explains on his faculty webpage that a video showing him shattering a student laptop that he has frozen in liquid nitrogen depicts violence done to a machine that was "non-functional before freezing" and that the putative owner is "a plant from a willing student accomplice." However, Mullen goes on to assert that the video does enhance his message about how students "in a large lecture class can be distracted by others watching movies, reading news-websites and playing games on their laptops." He also notes that he "had warned students many times about not using their laptops during lecture" and that his "demonstration did have the desired effect of improving student attention."[47] Another video in which a professor destroys a cell phone supposedly worth "several hundred dollars" to discipline a texting student ends with the student taking a bow. To minimize confusion, it is labeled as a skit on the official YouTube channel of Seattle Pacific University, where it was shot.[48] Much like popular "lip dub" videos shot with complex continuous shots at college campuses, the genre of the viral video about a destroyed student communication device often requires many confederates to manufacture the footage showing the supposed evidence of mutual disrespect.

If an authentic "angry professor" moment is captured in earnest, either by students with cell phone cameras in the audience or by tech support recording the class for posterity, such outbursts might be remixed, remediated, or parodied later. After Cornell lecturer Mark Talbert interrupted his lesson plan for two minutes to seek out and castigate the student who had emitted what he called an "overly loud yawn," his faculty temper tantrum was watched by over seven hundred thousand people on YouTube.[49] In the video, as Talbert insists on maintaining the boundary between "informal and impolite," he makes theatrically menacing comments like "my bad side is as bad as my pleasant side is pleasant," hunts for the offender among the rows of anxious students armed only with their electronic clickers, and finally yells that the guilty party should "walk the hell out." Soon after the video was posted, Talbert became the subject of ridicule with themed T-shirts, a Facebook-organized "yawn in" to protest classroom conditions, and a number of dance remixes featuring digital modulated and synthesized versions of Talbert's eruption with "my bad side is as bad as my pleasant side is pleasant" serving as a refrain.

Some angry professor incidents do prove to have physical and legal consequences for faculty who lose their cool. At Valdosta State University, Professor Frank Rybicki was charged with battery after closing a laptop on the hands of a twenty-two-year-old female student who refused to go offline during his Mass Media class. Rybicki had accused her of visiting noncourse-related websites, and an argument had ensued. Students who came to Rybicki's next class meeting were greeted by police officers who had come to interview witnesses and warned against discussing the case on Twitter or Facebook.[50] Yet generational allegiance seemed to do little to shape loyalties in the Rybicki case. In online comments many students at VSU expressed support for their professor and contempt for their classmate.[51] Over a hundred students joined the "Team Rybicki" Facebook group that was created to rally others to his cause and to organize protests over his treatment by authorities.[52] A number of youthful supporters wore "Team Rybicki" T-shirts to his trial.[53] In an interview after his acquittal, Rybicki described how when he "went to put the laptop down, she tried to pull it away from me, and she claimed that her finger got stuck." Rybicki "didn't see the finger get stuck" and argued successfully at trial that he had no intention of causing her any physical harm.[54] Although Rybicki argued that "the real issue in the case was the right of a professor to maintain the classroom as a learning environment," university administrators were still planning to terminate his contract and end his tenure-track employment as a result of the altercation.[55]

In other cases it has been the professor who was the one to bring suit. Part-time chemistry professor Kenneth Annan sued his supervisor at City College of Chicago for defamation; Annan claimed he was humiliated by security officers before being arrested for battery after confronting a student who refused to stop filming him with her cell phone.[56] Annan argued that cell phone batteries were a hazard if Bunsen burners were in use and that the offending student was not only disrupting his lecture but also endangering fellow students with her use of the ubiquitous communication technology.

Online videos in which students confront teachers in classrooms with multiple devices for recording and display can also generate online drama as viewers try to identify the guilty and innocent parties from what can be a confusing, multiperspective archive of user-generated moving images. For example, student affairs officials at a special "behavior of concern" workshop at my university showed footage from a Florida Atlantic University biology classroom in which a deranged undergraduate seemed to be threatening to kill her Asian-American professor and attacking a mixed-race cohort of her fellow students in a tirade focused on asking why "evolution kills black people."[57] The video shows multiple students with cell phones simultaneously recording the event rather than containing the chaos.

After assaulting those who were attempting to remove her from the room, campus police used a taser upon the student, which made the incident recognizable to those who had watched another genre of video about college-authority dynamics: the video in which a student is ejected from a public place and tasered by police. Two of the most famous videos of this genre are those depicting the 2006 incident involving UCLA student Mostafa Tabatabainejad being removed from Powell Library and the University of Florida incident that turned Andrew Meyer into a minor celebrity in 2007 after he disrupted a John Kerry campaign rally. Both videos were watched by millions, and the Meyer video turned into a bona fide meme that inspired multiple remixes of his catchphrase "Don't Tase Me Bro" vying for attention online.

Sympathetic media outlets pointed out that the student in the Florida Atlantic University video, Jonatha Carr, was upset about the recent shooting of unarmed black teenager Trayvon Martin in another Florida community[58] and argued that her outburst was much more comprehensible in this context. However, some African-American vloggers weighed in against her. One opined: "This is not where you go to perform; this is where you go to learn. ... You make participants in affirmative action look really bad now."[59] Racial stereotypes often are perpetuated by such in-classroom videos. Unfortunately, the search term "black student" in YouTube quickly brings up videos of combative students among the top results, such as "Student Punches Professor," "Black Student and Teacher Fight," and "Asian Teacher Black Student Argue Over Food Stamps."

Viral videos that show students and teachers in conflict may receive hundreds of thousands of views. They depict educational interactions gone wrong in a variety of classroom settings, from grade school to graduate school, and have become a truly global genre with examples represented on YouTube in the many languages in which the theater of conflict takes place. "Don't Mess with This Teacher" stars an indefatigable Thai female teacher, wearing a yellow satin dress and pearls, who interrupts her PowerPoint presentation to break a student's ringing cell phone and then resumes lecturing.[60] Sometimes these videos are translated to give more information and provide context that counters misinformation that emerges after viral dissemination has taken place. In "Prof Smashes Laptop: What REALLY Happened," a French language video annotated in English, the additional text informs the YouTube audience not only that the laptop being smashed belongs to an intrusive outside evaluator rather than to an obstreperous student, but also that the entire event is an elaborate April Fool's joke—even the supposed evaluator is an actor in the scene.[61] Although students have laptops open during the class, they do not appear to be egregiously Web surfing, and they clap in support of their professor after he seems to lose his cool and gasp in amazement when he reveals his entertaining ruse.

Taking Digital Rhetoric Seriously

After being involved with digital media and learning in various forms of public service in higher education for over two decades, I must admit that it is sometimes difficult to ignore the presence of a vocal minority of "abolitionists" who are determined to end institutions rather than mend them. On mailing lists these extreme true believers in a leveling digital revolution will declare that the university is the problem rather than the solution, or that students learn in spite of their professors rather than because of them. As a classroom veteran, I have seen how first-generation college students from families without a history of higher education sometimes need the most human and humane forms of live, in-person instruction. To not be discouraged, these students require pedagogical responses that are more fault-tolerant than either automated systems or peer groups dealing out the Internet's toughest forms of tough love. I am not sure that self-learning at home at the computer is the right approach for every subject or every student, and I am suspicious of how for-profit educational businesses may be driving the debate while disregarding the cost to community colleges and other traditional local institutions.

Furthermore, we need to share solutions for dealing with the problem of bad actors at the keyboard who may participate inappropriately or not at all when the boundaries of the classroom and the new social roles of educators are not clearly defined. I would urge optimists who celebrate informal culture uncritically to acknowledge the presence of griefers, spammers, astroturfers, phishers, stalkers, lurkers, and flamers in their virtual learning environments and to examine their own wishful thinking about corporate partners from high-tech industries who may be entering education with only a profit motive at work. On the other hand, I would also urge the pessimists, who are often armchair theorists, to maintain less critical distance and to pitch in more with actual students, parents, and citizens—thereby adding their muscle, as well as their prestige, to digital initiative efforts. Unfortunately, forms of academic labor that are devoted to building digital initiatives benefiting more than one university or geographical location often count as service rather than research, or they are seen as "community outreach" rather than serious scholarship.

If we are to ever gain ground in the digital university, we all need to question our own myths and understand our own inheritances; continuing to either ceaselessly invent and celebrate the new or demolish the half-measures and compromises of others is only causing us to lose ground with both our students and the public.

2 The War on Learning

In 2010, I visited a course designed to teach digital literacy skills to college freshman that was being held in three rooms: One offered a "live" faculty member holding forth on the stage, while the other rooms presented a broadcast version of her performance. It was a required general education course that focused on teaching Alice, an object-oriented computer programming language with a simple, graphical user interface designed for K–12 students. I was skeptical about whether Alice was the right programming language to teach college students, but I knew that the lecturing faculty member had a good reputation for innovative teaching and deploying student-centered learning techniques.

At the beginning of the quarter, students in all three rooms had purchased wireless handheld clickers that could monitor their class attendance, assure that their attention did not wander too far from the subject matter during class time, and—hopefully—measure their performance at reaching the course's stated learning goals. During my visit to the class, I soon noticed that, because they had to select the best answer quickly among several plausible choices as a digital clock counted down, students perked up each time the screen displayed multiple-choice questions about computer programming techniques. Not all of the questions on the screen were straightforward, problem-solving questions about choosing the right line of code or the right order or the right operation: sometimes the questions involved fundamental definitions or more abstract reasoning about choosing an approach. To encourage collaborative learning, students could also periodically confer with each other in small groups and debate the merits of the different answers.

In the clicker room with the live performance by the professor, I watched about forty-five minutes of a college instructor trying to do everything right. In addition to the clicker quizzes and small group discussion, she encouraged active learning by soliciting questions and comments from students using a microphone. Despite her best efforts, about half of the students in this lecture hall were partially engaged; they only paid attention during the quiz portions of the session. Much of the time, these students seemed to be considerably more interested in non-class-related activities, such as games and social network sites, taking place on laptops and cell phones. The other half behaved more attentively, like students who could see the professor and, in turn, be seen. Occasionally there were small dramas in the lecture hall, such as when a student could not get his or her clicker to function, and there would be a flurry of desperate activity around supplying batteries for a dead clicker from clickers that had already clicked in before the ticking clock ran out of time.

I then went to the large lecture hall to the immediate right of the live lecture, where no professor was actually present. It was part of an initiative intended to pilot a distance-learning course that would teach computer programming to advanced placement computer science students who might not have such courses offered at their high schools, which was being tested on first-year college students at my institution. No one was at the podium in this lecture hall. The faculty lecturer appeared only as a fuzzy apparition moving around on the right-hand screen. The room was brightly lit, like the first lecture hall, but, even though both the content delivered and the required clicker quizzes were identical to those taking place in the room next door, students were much more disengaged from the spectral lecturer and her phantom interactive activities. Almost all the students in this class were texting and Web surfing, with only brief pauses to answer quiz questions with their clickers. Some of these students were engaged in highly inappropriate behavior: students applied makeup, talked on cell phones, and held conversations at regular volume with each other about matters unrelated to class. In this lecture hall, small group discussions quickly strayed into non-class-related topics and continued long after the video broadcast instructor resumed lecturing. I asked a graduate student teaching assistant patrolling the aisles with a microphone if students were ever discouraged from pursuing such non-class-related activities, and he said, "Never."

Sitting in that lecture hall, I felt as though I were watching an experiment in digital learning that had gone terribly wrong, but no one seemed willing to tinker with the experimental protocol or question the basic assumptions about learning and teaching that were dramatized. In that lecture hall, the once-covert war between "us" and "them" had turned into an open battle between "our" technologies and "their" technologies. On one side, we faculty could reset countdown clocks and forgive clicker mishaps, but they had access to wireless networks and mobile telephony. My sensibilities as an educator were assaulted by this spectacle of mutual contempt, but it seemed futile to try to engage anyone present in a conversation about what they were perpetuating and why.

For the last fifteen minutes of my observation, I took refuge in a much smaller room, which was configured like a large classroom, rather than a lecture hall, and was located to the left of the live lecture. This room was dark and the broadcast instructor was projected with close-up shots on a relatively large screen, where students could clearly see the professor's facial expressions in the intimacy of the room. While viewing the same instructional content, this group was much more attentive than the group in the larger, brighter space, and behaved as if at a movie theater with reverent silence. I was dispirited to see that there was also less interaction during the discussion portions. This room, however, was standing-room-only. There was a kind of peace to be found in the passive darkness, a respite from the war on learning being staged under bright lights a few doors down.

The Twitter Experiment

A few weeks later I had to struggle with my own scene of pedagogical failure. This took place in a large lecture hall in the first iteration of an upper-division lecture course of 230 students that I had scaled up from what was once a small senior seminar with a tenth the number of students. The course had formerly been taught around a table in the cozy setting of a nontraditional "smart" classroom equipped with multiple screens, individual laptop hookups, and tablets to encourage active learning and, since all Web surfing activity was automatically shown to the group, personal accountability. The new course was held in a large lecture hall in a campus conference center, where it was difficult even to see my students' faces.

From the start of the quarter, I was worried about the impersonal nature of this large class configuration. I tend to prefer lecturing to audiences with about half as many students so that I can learn most of their names. After all, I knew that Dunbar's number, 150, supposedly represented the upper limit of a functional community size defined by manageable social networks. According to the Dunbarians, primates with successively larger brains supposedly evolved with a cognitive limit on the total number of individuals with whom they could maintain stable social relationships.[1] However, many required courses in public universities are taught in lecture halls with seating capacities far over that number. Given the problems of scale, I supposed that students' desire to rebel against such factory-scale teaching might be understandable.

At two o'clock on January 27, 2011, my pedagogical experiment began. In an effort to get more of the two-hundred-plus students to actively participate in the class discussion that day, I chose a hashtag on Twitter, #cat125, and invited a roomful of otherwise silent and passive UC San Diego seniors in the Culture, Art, and Technology (CAT) program to tweet for thirty minutes and, thus, add their comments to a freeform, evolving stream of text in the public record. Two weeks earlier, I had posted the first message: "Soon the experiment begins #cat125." By using Twitter, I had hoped to create an environment of lively exchange about course content. In the days leading up to the experiment, I cheerfully projected the #cat125 Twitter feed in class and walked the undergraduate audience through Twitter's basic lexicon of URL-shorteners, hashtags, and "@" call signs. Fellow academics who had followed my tweeting at conferences or professional conventions hailed the effort. When my tweet "I am showing my #cat125 class how to use Twitter now" appeared on her Twitter news feed, game scholar Nina Huntemann sent "a hello wave from Boston" to my students. Open publishing advocate Kathleen Fitzpatrick soon chimed in too: "Hi, #cat125!" Cheryl Ball at Illinois State observed that she would soon be doing the same with her "multimodal seminar students." Katherine Harris at San Jose State even posted an exuberant "Well, hello everyone! We're listening; jump into the conversation at any time #cat125." Meanwhile, students were tentatively testing the waters with first messages like this one: "@lizlosh this is my first time using twitter, and my first tweet ever. Hello #cat125."

Despite these auspicious beginnings, I knew that pedagogical disaster was always a distinct possibility. Social media maven danah boyd had described her own Twitter debacle that took place during a talk in November of 2009 in which "outbursts and laughter" revealed that Twitter "had become the center of attention, not the speaker."[2] As one of boyd's online catcallers later explained, "the more subtle the speaker's point, the more impatient and nasty the audience became," as the wall of comments "made a spectacle of the crowd's impatience and anxiety feeding on the speaker's inability to respond."[3] He argued that the use of Twitter "united us not as a single group receiving challenging ideas from a thoughtful orator but as quite separate individuals struggling to listen, read, respond, and make sense of the event."

Certainly, as students chose screen names, there were already signs that they wanted to remain anonymous and ironically distant from the proceedings: their online monikers included pseudonyms such as "cat125isthebest,"

"IHeartCat125," "CATmademetweet," and "ChancellorFox," which was also the name of the head of the university at the time. However, because witty, short-form communication that is heavy on humor and light on gravitas is often the preferred vernacular for Twitter, I lauded rather than chastised them for their inventiveness.

When the rehearsals were over, and the time for mass performance came, the students' collective broadcasting about their own inattention quickly dominated the channel. As the experiment commenced in the classroom, multitasking and divided attention quickly became major themes in the Twitter stream. One student admitted "I feel guilty to have made this twitter account during class … but I blame new media for my A.D.D." Others focused on their sensory perceptions of stimuli unrelated to the lecture with comments like "I just noticed all the flags on the ceiling" or "Who is eating tasty delicious smelling food? What delicious vittles are you enjoying?"

During a class session that was intended to be devoted to disability videos on YouTube, even comments somewhat germane to the clips that were shown emphasized the students' difficulties with sustained attention and engagement—particularly if the video timeline played for more than a few minutes. Some credited their discomfort with the material as a reason to tune out. As one student opined while a video created by an autistic person played, "i feel alienated when I was watching that video earlier today. I had to stop it in the middle."

Soon the Twitter stream moved to themes of collective action rather than individual perception. Students leading this transition were actually paying much more attention to the popularity metrics that drive social media platforms like Twitter than to the course's content during the experiment. One wrote, "I wonder if we could actually get #cat125 to become a top trending topic. that would be ridiculously fun." Another student bemoaned the fact that "clearly we aren't tweeting enough if we aren't even a top trending topic." Soon notices informed them that they had reached their collective goal and that #cat125 was trending in San Diego.

Within minutes, students also made plans for coordinated counterprogramming to my carefully orchestrated presentation about disability and social media in honor of Diversity Day on the campus. First, there was the suggestion to all drop pencils at a given time. Then, a particularly confident user of Twitter and other forms of social media gave a command that was retweeted throughout the room: "i say at 2:30 we all randomly start coughing … the proff isn't tweeting so she won't know whts going on hahaha." I had, in fact, seen the pencil drop tweet and was aware that there was dissatisfaction with both the content of the class and the experiment being expressed in the room, but I had intended to pursue nonintervention to show respect for their peer-to-peer interactions. I was also preoccupied with the special guest who had just arrived in the lecture hall, Professor Tom Humphries, associate director of the Education Studies program on my campus and coauthor of the influential books *Deaf in America: Voices from a Culture* (1988) and *Inside Deaf Culture* (2005), who had come to give a guest lecture about deaf culture and YouTube for the second part of the class.

Right on schedule at 2:30, the coughing began in the middle of a video created by a schizophrenic female journalist explaining her disability to a broad online public. Tweets such as "hahahah at the coughing," "hehe," and "halls anyone?" soon appeared with the #cat125 hashtag. In honor of the presence of Professor Humphries, one student suggested that "We should 'applaud' in American Sign Language instead." Another posted "Say when." But the coughing and laughing continued until I announced the end of the use of laptops and other mobile devices and introduced the Professor Humphries to the group. Among the final words registered in the stream as students were closing laptops and cell phones was the student who wrote this plaintive farewell: "I didn't know what to say … #cat125 deleting account now …" In expressing both her disapproval of the conduct of her classmates and her own form of ironic distance from the experiment, another wrote, "I remember that from eighth grade. #cat125. This is why we can't have nice things!"

When I have told the story of the Twitter experiment and the coughing exploit that followed to my academic colleagues, I have heard many generalizations to explain the students' seemingly callous behavior in front of me, my guest, his sign language interpreter, her class of student-interpreters, and a deaf peer seated in the front row on a day designated for tolerance and understanding and intended to highlight how new media could remedy inequities in race, class, gender, sexuality, and ability. Some instructors chalked it up to students' impatience or fatigue with what they might see as a politically correct lesson devoted to liberal moralism rather than useful knowledge that could advance their academic careers. Other faculty—who expressed outrage or resignation or both in response to my tale—told me that such antics were symptomatic of a broader undermining of respect for knowledge among the digital generation and a waning of the traditions of sustained attention once fostered by print culture. The worst reactionaries in this group even suggested, in a live-by-the-sword/die-by-the-sword logic, that all classes in and about the Internet would necessarily devolve in this way, so academic departments should think carefully before placing classes like my CAT 125 on their schedules.[4] In contrast, there were others, more

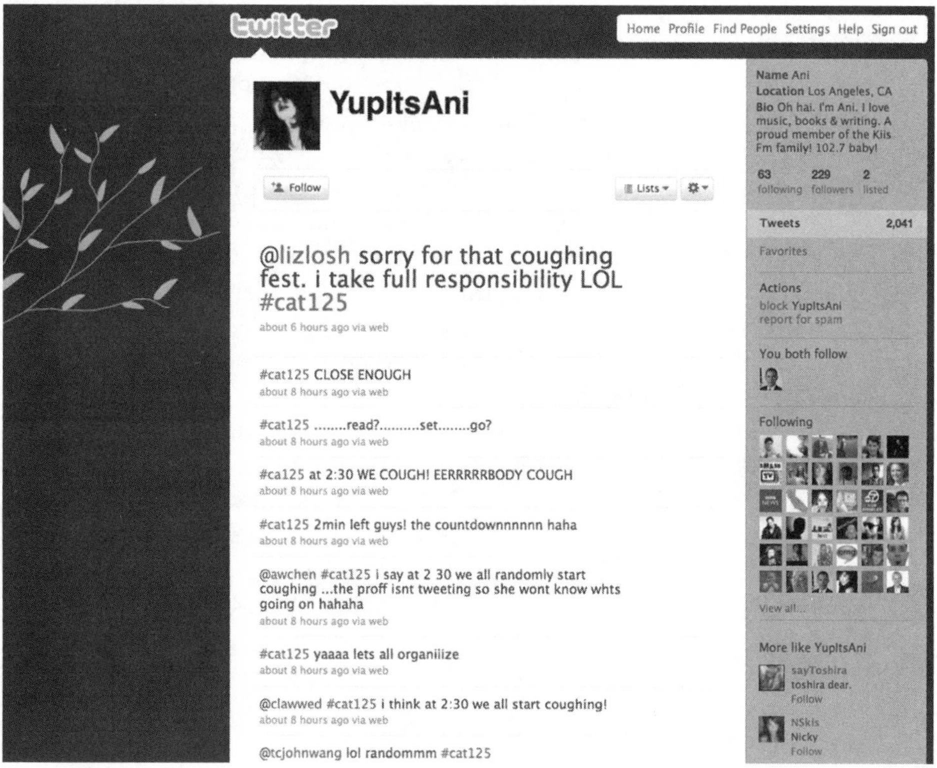

Figure 2.1
CAT 125 student's Twitter feed. Courtesy of Ani Stepanian.

sympathetic to the students, who credited the taunting to the numbing effect of communicating at a distance without the social glue of face-to-face communication, but they also blamed online technologies for the degraded state of mutual alienation in the lecture hall.

Nonetheless, many of the students saw revolution rather than devolution at work when recalling the events of that day. They understood the coordinated coughing stunt as a manifestation of the power of self-organizing groups equipped with distributed communication tools. On the class blog, a number of students made analogies between the Twitter-orchestrated coughing exploit and the 2011 "Arab Spring" (which was taking place at about the same historical moment as our CAT class) that harnessed the activities of flash mobs using mobile devices to help topple dictatorial governments in the Middle East and North Africa. From the students' perspective, a number of them saw their subversive uses of the technology as an act of political power, analogous to the smart mobs using cell phones, text messages, and social networks to destabilize authoritarian regimes.[5] One student blog post argued that the "humorous/disruptive scene" that resulted from the Twitter experiment is "not something to be belittled."[6] Another blog entry described the instigator of the coughing as "one clever student" who led the others to "feel quite proud of ourselves."[7] The "clever student" who led the in-class revolt actually produced a video essay that promoted the analogy further and argued that the Twitter spontaneous-organization exercise made power structures in the lecture hall more visible to the students.

For the first seventy-nine seconds of her video, the ringleader of the coughing exploit retold the story of how street protests against Egyptian President Hosni Mubarak were spurred by microblogging and social network sites. She illustrated her narrative of carnage and triumph with stills and footage shot in Cairo that she had collected from the Internet.[8] According to this student, the same principle of organizing civil disobedience through lateral channels applied to a college lecture hall where "some students were confused, and some were distracted, but others saw the potential to organize."[9] She insisted that the exploit "was my way of testing the limits of this social experiment and demonstrating our presence as students," and its success confirmed the "speed, power, and overall ease of organization on Twitter."[10] From her perspective, the Twitter experiment was a rousing success, at least from the point of view of once-disempowered students who had found an ability to take collective action through the microblogging platform.

After the Twitter experiment was over, I got a number of substantive suggestions from students for how the course could be better the next time and how the experiment could have succeeded from an instructional perspective. Expressed in these postings was a desire for students' common digital efforts to be directed toward specific and measurable learning goals rather than to the flash-mob behavior celebrated by others. The graduate student teaching assistants who led discussion sections for the CAT 125 course were encouraged to publish articles and present at scholarly conferences about their own pedagogical insights from this incident and to share what could be learned from this particular teaching failure story with the broader teaching community.

Certainly, the Twitter experiment helped students understand that digital tools could be used politically on campuses, but it was clear that, once one particular aspect of unequal power relations was revealed, those undergraduates still needed theoretical apparatuses and practical guidelines for

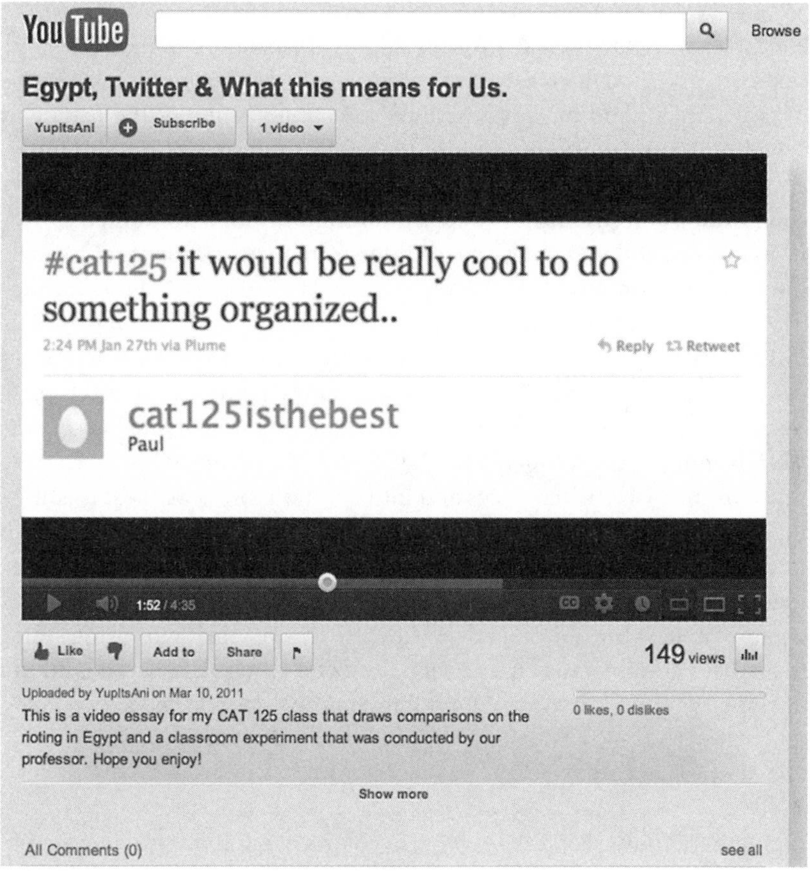

Figure 2.2
CAT 125 student's video about the incident. Courtesy of Ani Stepanian.

more meaningful ethical, rhetorical, and historical reflection. I'm not sure I can accept the easy analogy that the students who pointed to the Egyptian "Twitter revolution" were making between radically different political and cultural contexts or the sweeping rhetoric that dismissed lecture-hall teaching as dictatorial in nature and justifiably deposed, but I did validate the legitimacy of their viewpoints by urging them to be included in an issue of an online journal, the *New Everyday*, that was devoted to the current distraction panic and the possibilities of productive distraction created by student-centered social media. By collaborating with the ringleader, I hoped to make lemonade out of lemons and understand the incident as a set of usable lessons learned.

As editors Brian Goldfarb and Alexandra Juhasz explained, their journal issue would "include actual media practices to dispel the common diagnosis that networked media is rewiring young minds, displacing valuable forms of engagement, and making sustained reflection a thing of the past" and "challenge pathologizing discourses that frame young users of social media as simultaneously victims and threat (the teen who doesn't listen, who won't concentrate, who will fail to learn)."[11] Juhasz and Goldfarb wanted to provide nuanced counterarguments to mainstream media accounts of the new information economy and to avoid hyperbolic terms, such as "epidemic" or "death," that dominate the news. Furthermore, Goldfarb and Juhasz insisted "that adults and youth can work together to make better sense and use of the interruptions, disturbances, and amusements introduced by new media," but they also admitted that "multitasking, fragmented and interlaced communication fostered by the culture of new media" are occurring "even in the places where focus and introspection might be most useful."[12] Juhasz and Goldfarb, like many others described in this chapter, attempt to navigate a middle path of practical engagement and thoughtful resistance as early adopters of technology who also raise questions about its appropriateness for learning environments. This pragmatic bridging seems to many to be desperately needed at a time when the gulf between a culture of knowledge and a culture of information is only getting wider.

The Rhetoric of Crisis

It's not hard to see that, within popular commentary about the successes or failures of education in our culture, a rhetoric of crisis has been at work for at least a century; variants of "Why Johnny Can't Read" stories seem to grab headlines every decade. Many of the same people involved in the "death of poetry" debates of the 1990s that I wrote about in my graduate school dissertation[13] are now, in the twenty-first century, bemoaning the death of reading itself. For example, former National Endowment for the Arts Chairman Dana Gioia, who wrote the book *Can Poetry Matter?*, describes his agency's report, *Reading at Risk*, as a "bleak assessment" to be read with "grave concern" as it documents a "dire" situation of "imminent cultural crisis" that is demonstrated factually by exhaustive surveys showing steep declines, particularly among the youngest Americans, in time spent reading literature.[14] Gioia blames diminishing attention to reading, and the declining test scores of students faced with traditional print, on a "massive shift toward electronic media for entertainment and information" that are assumed to require less formal training and education to consume and to be closer to the preliterate products of "oral culture."[15]

Mark Bauerlein, one of the *Reading at Risk* report's coauthors, provides an even more negative view of the young in *The Dumbest Generation: How the Digital Age Stupefies Young Americans and Jeopardizes Our Future.* As he did in *Reading at Risk,* Bauerlein buttresses his claims with pages of statistics about falling reading scores and youth apathy toward maintaining consistent reading habits. But as one *Newsweek* reviewer notes, Bauerlein's complaints about the ignorance of the young are hardly novel in the literary record or strongly correlated to the rise of digital technology.[16] After all, defending established culture against the ignorance of youth is a rhetorical trope that goes back to antiquity. Furthermore, as *Reading at Risk* itself acknowledges, acquiring the ability to read deeply is not just a linear process of universal acquisition that is based on maturity; our attitudes about reading have rich cultural associations, and assumptions about "good" readers and "bad" readers may be shaped by race, class, and gender,[17] as well as by age.

It's also worth noting that Bauerlein's diatribes are not restricted to castigating Internet use; he also rails against libraries for providing popular services such as DVD check out and computer access and complains about educational reformers who argue for the value of "active learning" in popular initiatives for problem-based, project-based, or inquiry-based curricula.[18] The custodians of challenging and difficult forms of knowledge acquisition supposedly cheapen traditional literacy by making it more approachable and accessible, so that Bauerlein often comes off as a prohibitionist and a puritan, eager to forbid people's everyday digital practices.[19]

Contemporary politics shows us that this kind of cultural fundamentalism can succeed, particularly when the press demonizes file sharing, the use of social networks, and video game play—but what is the cost to the free exchange of ideas? In his section on "culture warriors," Bauerlein seems to grant the importance of being able to argue for both sides, but he himself seems unwilling to acknowledge counterarguments from those who don't subscribe to his deficit model of intellectual decline. Many of the researchers missing from Bauerlein's bibliography are also known for their use of in-depth ethnographic techniques that involve observing subjects for long periods of time and developing rich case studies around the behavior of particular young computer users. It is significant that these researchers often don't paint an entirely rosy picture of the digital literacy of the supposed digital generation.[20]

Granted, Bauerlein may be right that there are many ways in which Facebook and MySpace can extend the mindless banality of high school groupthink into the broader culture. Anyone who has gotten a friend request from a former grade-school tormentor knows that the pointless popularity contests of preadolescence can continue in such social network sites. When she talks about the preeminence of gossip, Mimi Ito admits as much in her much-cited report, completed for the MacArthur Foundation, with the colloquial title *Hanging Out, Messing Around, and Geeking Out: Kids Living and Learning with New Media*, which is often abbreviated to HOMAGO among those who have adopted Ito's ideas. Although Bauerlein argues that championing "enhanced connectivity" and encouraging the "indulgence of teachers and journalists" only feeds the "adolescent vice" of "peer-absorption,"[21] Ito insists that facilitating peer-to-peer learning has always been an essential part of building effective learning communities.

As someone who once taught in a traditional humanities core program with a great books curriculum, I tend to agree with Ito. The canon is full of stories of peer-to-peer learning. For example, much of the sixteenth-century autobiography of Saint Ignatius of Loyola is devoted to his interactions with his fellow students, and even fourth-century Saint Augustine—who deeply admires enlightened superiors such as his teacher Saint Ambrose—describes gaining significant insights in his education from his peers. Certainly, any classroom instructor will tell you that learning environments are only enhanced by the presence of other engaged learners in the room.[22]

In *HOMAGO*, Ito describes two basic kinds of youthful Internet users: (1) those who use social computing technologies to facilitate high school affinity practices, such as gossip and status-checking, and (2) those who seek out more advanced forms of expertise in specialized knowledge communities.[23] Bauerlein totally dismisses the latter group in his attack on the former; he focuses exclusively on how the Internet spreads the banalities of top forty and tabloid culture rather than addressing how it also disseminates information from a "long tail" of more eclectic and specialized interests[24] that make the print resources of even the most specialized and arcane rare-books library accessible to all with an online connection. Instead of talking about strictly defined digital generations, Ito focuses on media ecologies in which parents, teachers, and other adult mentors may play a significant role in "geeking out" activities.

The type of geeking out that Ito describes in *HOMAGO* is also important in the essay (later a book)[25] "Hackers and Painters" in which Internet guru Paul Graham claims that the problem with high school is not the cultural power of the young but rather their impotence within adult society. Graham argues that without the apprenticeship systems of traditional societies, teenagers are locked out of meaningful participation in adult culture, and those who try to earn membership among the grown-ups by showing off their intellectual gifts or trying to be productive individuals by, for instance, mastering an instrument, building robots, programming computers, or playing virtuoso games of chess are often tormented by their peers as nerds. Access to computing resources not only strengthens such nerds' abilities; it also allows them to see that they are not alone in their quest for adult acceptance and membership in society.

In *The Dumbest Generation*, Bauerlein talks about how Frederick Douglass, W. E. B. Du Bois, John Stuart Mill, and Walt Whitman were liberated by books, but strangely, Bauerlein does not seem to realize that for some liberation comes from access to the computer rather than the printed page. For example, danah boyd has written about the importance of the Internet for gay teenagers in the poisonously homophobic atmosphere of high school, in which the ultramasculine jock and the ultrafeminine homecoming queen/cheerleader rule as the royalty on campus.[26] The possibility that the Internet could facilitate potentialities for genuine—as opposed to fatuous—"resistance identities,"[27] to use Manuel Castells's terms, is never seriously entertained in Bauerlein's book; nor does he address the ways that some Internet practices can serve as correctives to teen conformity and consumerism.

There is less overt age discrimination and a much more muted tone of hysteria in *The Shallows: What the Internet Is Doing to Our Brains* by Nicholas Carr, but many have read it alongside Bauerlein's work as another data-rich bestseller confirming the national decline in traditional reading skills. However, Carr is much more guarded in adopting the rhetoric of panic about media influence and approaches his subject with more historical perspective. Not only is Carr better versed than Bauerlein in the sophisticated media theories articulated by Marshall McLuhan and Walter Ong, he also reminds his readers that his anxieties are old anxieties and that Plato once wrote about the dangers to memory posed by the introduction of written texts. Furthermore, unlike Bauerlein, who situates himself as a smart reader able to resist the dumbing powers of the Internet, Carr opens *The Shallows* by admitting his own difficulties with devoting his attention to sustained and deep reading the more he uses the Internet to conduct research:

> Whether I'm online or not, my mind now expects to take in information the way the Net distributes it: in a swiftly moving stream of particles. Once I was a scuba diver in the sea of words. Now I zip along the surface like a guy on a Jet Ski.[28]

Carr narrates a personal tale of digital distraction in the beginning of the book, which he returns to near the end when he explains how moving from the wired suburbs of Boston to the mountains of Colorado actually allowed him to finish his writing project.

Carr can be persuasive as he describes the changes that he observes in his own reading habits—changes that he attributes to the plasticity of his own brain and explains through contemporary neuroscience. Certainly, research on brain plasticity in recent years has produced startling results; apparently, even brains that have had an entire hemisphere removed can compensate for this catastrophic loss with remarkable agility.[29] According to Carr, environmental factors and certain kinds of interactions with information technologies can actually "rewire" the brain's electrochemical pathways; this can occur even if the adult brain once had seemed firmly hardwired for deep reading by educational experiences in childhood. In a rapid march through a history of several millennia, Carr details how brains were changed by primitive tools, by maps and clocks, by writing and print, and, eventually, by the advent of digital computers. Such tools not only extend human cognition, Carr asserts; they also change cognition fundamentally. Carr's emphasis on nurture rather than nature might seem to offer more hope to educators than Bauerlein's diatribe, but the forms of biological and technological determinism that Carr observes seem more difficult to reverse than Bauerlein's social coddling.

One of the experts that Carr cites as a witness to the decline of book literacy has articulated some significant objections to Carr's arguments, based on her own analysis of how he uses results from neuroscience experiments. Although English professor N. Katherine Hayles agrees with the technogenesis hypothesis generally and points to the telegraph as her own example of a technology capable of rewiring the brain, she takes issue with the way Carr uses scientific evidence to shut down debate and asks if his pessimism is really tenable. In defending the possibilities of newer forms of hyper-reading to enrich knowledge gained through traditional close reading, Hayles also criticizes her colleagues in the humanities:

> At this point, scholars in literary studies should be jumping on their desks and waving their hands in the air, saying "Hey! Look at us! We know how to read *really* well, and we know how to teach students to read. There's a national crisis in reading? We can help." Yet there is little evidence that the profession of literary studies has made a significant difference in the national picture, including on the college level, where reading abilities continue to decline even into graduate school. This is strange. The inability to address the crisis successfully no doubt has multiple causes, but one in particular strikes me as vitally important. While literary studies continues to teach close reading to students, it does less well in exploiting the trend toward the digital. Students read incessantly in digital media and write in it as well, but only infrequently are they encouraged to do so in literature classes or in environments that encourage the transfer of print reading abilities to digital and vice versa. The two

tracks, print and digital, run side by side, but messages from either track do not leap across to the other side.[30]

Hayles insists, following L. S. Vygotsky, that professors need to be attuned to the "zone of proximal development"[31] if they are to bridge the gap between their own close-reading practices and their students' online hyperreading practices. Furthermore, she asserts that forms of hyperattention may have evolutionary and cultural benefits and should not always be disparaged by instructors. Like Goldfarb and Juhasz, Hayles attempts to negotiate the ground between distributed and focused forms of attention and to lessen investments in the superiority of one supposed generation over another.

Unfortunately, such calls for moderation in implementation and balanced efforts are far too often cast as minority opinions—ones that don't get into the pages of bestselling books. Instead, as Deborah Holdstein argues, even university administrators are getting drawn into what she calls the "Paradoxes of 'Crisis' and 'Excellence'"[32] that dominate discussions about the university's future. Holdstein writes largely about the perceived "crisis in writing" rather than the "crisis in reading," but she argues that universities are in a perpetual state of whiplash as they seek to either trumpet their superiority or warn about the dangers of imminent disaster. Both strategies are designed to gain financial resources from the public for needed educational services, although appeals with similar ends may take different forms. In reflecting the novelty-seeking and publicity-seeking tendencies of our broader society, universities prioritize narratives of individualism, moralistic rhetoric, simplistic instrumentalism, and the enforcement of norms—even if theorization and reflection are what is most needed.

As the obituary for reading is being written, so is the obituary for the university and for the forms of higher education that are tied to print culture. Often the death of the university is imagined as the death of the humanities, and it can sometimes be difficult to determine which death is supposed to kill the larger body of knowledge. A recent crop of books that includes *The Last Professors*, *The Marketplace of Ideas*, *Not For Profit*, and *Unmaking the Public University* has attempted to analyze what appears to be a crisis in the humanities and the imperiled status of the intellectual class within it. Often, instructional technology is part of that larger narrative of woe. For example, in *The Last Professors*, Frank Donoghue examines a cynicism about learning that dates back to the robber barons of the industrial gilded age; he sees a precedent for the current instructional-technology movement in a "1910 Carnegie Foundation report on academic and industrial efficiency" in which the author "maintained that it made business sense for course materials and lectures" to belong to the scholarly workplace and its daily operations rather than to independent faculty who refused to serve as an alienated labor force. Donoghue concludes that collusion between "major publishing houses and university administration" with the "online course-management industry" could enact the recommendations of the Carnegie report a century after it was written.[33] He believes that the rising ranks of cheap adjunct instructors, cost-slashing bottom-line accountants, and technocrats devoted to new intellectual property regimes that generate new revenue streams for universities might dethrone the last professors and that instructional technology is a large part of this story of decline.

In Toby Miller's provocative manifesto *Blow Up the Humanities*, he notes that the title could refer either to his incendiary desire to destroy his own field as it exists today or his to intention to lift up the humanities with a hopeful trial balloon. He opens by asserting that the humanities as it exists today is fundamentally bifurcated between two separate camps:

> There are two humanities in the United States. One is the humanities of fancy private universities, where the bourgeoisie and its favored subalterns are tutored in finishing school. I am naming this Humanities One, because it is venerable and powerful and tends to determine how the sector is discussed in public. The other is the humanities of everyday state schools, which focus more on job prospects. I am calling this Humanities Two. Humanities One dominates rhetorically. Humanities Two dominates numerically. The distinction between them, which is far from absolute but heuristically and statistically persuasive, places literature, history, and philosophy on one side and communication and media studies on the other.[34]

Miller makes a strong case for the continuing viability of the more agile, and less classist, Humanities Two. However, he warns that it is too closely allied with neoliberal interests, military imperialism, and the deskilling of the labor market; this is because Humanities Two college students are often being prepared to join the creative classes in computational media. Instead, he posits that a third humanities should serve as a more ethical substitute for the existing two choices—one that combines "political economy, textual analysis, ethnography, and environmental studies such that students learn the materiality of how meaning is made, conveyed, and discarded."[35]

Not all professors from "Humanities One" and traditional institutions have concluded that distance learning is necessarily an expression of evil incarnate. Many recognize that older nontraditional students, international students, and students from certain kinds of rural or urban geographies who may live in research deserts need higher education too. Digital technologies can extend the scope of community outreach programs and the effectiveness of progressive research universities with a strong sense of social mission. Even former Harvard president Derek Bok calls for candor about the shortcomings of conventional, on-site lecture-hall teaching, including on Ivy League campuses. He claims that online learning is capable of producing compelling outcomes, especially if faculty are active participants as early adopters and if universities don't expect cost cutting from their efforts. Bok argues that developing high-quality, labor-intensive courses that may blend online learning with face-to-face live instruction will be critical as we live through a time of potential rapid change in higher education. [36]

Revolt from Within

When Professor Mark C. Taylor committed heresy by arguing in the *New York Times* that we should "end the university as we know it" by terminating tenure and dissolving departments,[37] public response was enormous. Although many responded most strongly to Taylor's attack on the job security of educators or his utopian calls to replace conventional disciplinary schemes of organization in permanent departments with "zones of inquiry," such as "Mind, Body, Law, Information, Networks, Language, Space, Time, Media, Money, Life, and Water,"[38] and others reacted to his calls for more online education, few readers contested his insistence that a crisis was at hand.

In the comments section of Taylor's *New York Times* opinion piece, two specific analogies frequently were used: (1) the university is in a crisis similar to that of the health care system, or (2) the university is in a crisis similar to that of the newspaper industry. Those who medicalized the crisis in the university thought that specialization and perverse reward systems were driving up costs, but they also thought that the university couldn't be dismantled because—like the health care system—there were immediate and urgent needs for universal access to its services, particularly when basic literacy and advanced skills in information technologies were even more essential to workers in an age of modern, high-tech globalization. In contrast, those who saw the university as similar to a media outlet shaping the news focused on how new technologies were driving down costs for the university's competitors, who could now disseminate information more efficiently (and more democratically) than those behind ivy-fortified walls. Just as in the profession of journalism, resistance to change in the profession of teaching would be futile and would only lessen public trust. If government depends on an informed citizenry, then the costs to the public suggest that action in the form of radical reform shouldn't be further delayed.[39] According to these commenters, universities must embrace change or face extinction as a result of their own irrelevance.

In the case of both health care and the newspaper industry, of course, the institutional character of their public enterprises contributes nothing to their basic mission. Few would argue in favor of further institutionalizing either health care or the newspaper industry, but institutionalization is seen as vital to the university and central to its principles of tenure, academic freedom, and peer review. Thus, one of the remarkable things about Cathy N. Davidson and David Theo Goldberg's *The Future of Learning Institutions in the Digital Age* is its willingness to tackle the "mythology of technology that its virtues, vitality, and values are 'free.'"[40] As leaders of the Humanities, Arts, Science, and Technology Alliance and Collaboratory (HASTAC) and the University of California Humanities Research Institute, Davidson and Goldberg well understand the importance of institutional infrastructure. Although other futuristic books tend to treat the term "institution" (and even words such as "education" and "teaching") as contaminated, favoring the term "learning," Davidson and Goldberg emphasize the fact that digital learning requires labor, capital, material resources, and organizational structures aimed at public reputation and sustainability. Unlike end-the-university manifestos that reject institutional frameworks entirely, the authors remind readers that "our traditional institutions" in higher education are notable for their "endurance and stability."

There are notable absences, however, in Davidson and Goldberg's initial master plan. Although the book claims that its "primary focus is higher education," it is remarkably light on specifics drawn from undergraduate education. Its "Portfolio of Virtual Learning Institutions: Models, Experiments, and Examples to Learn and Build On" contains almost no examples of work done for college credit; instead it presents K–12 initiatives such as the Gamelab Institute of Play, Quest to Learn, the New York City Museum School, and the School of the Future in Philadelphia. And in the transformation that they project in "modes of organization, structures of knowledge, and the relationships between and among groups of students, faculty, and others across campus or around the world,"[41] the "others" are largely constituted as off-campus participants who might be "strangers" who could "remain anonymous"[42] as they participate in life-long learning, contribute to wikis, serve as audiences for faculty blogs, and enter into symbiotic information-exchange relationships with those on campus remotely. Absent are the librarians, instructional technology specialists, and other members of non-tenure-track academic underclasses, who contribute much to institutional initiatives for digital learning but may be further marginalized by existing reward systems.

Because *The Future of Learning Institutions in the Digital Age* was vetted through a process of public online comments facilitated by the open-source CommentPress software as a platform for open peer review, Davidson and Goldberg describe it as serving as a novel "writing exercise" that challenges conventional academic publishing norms. Around the same time their book was published, two other notable books about embracing digital media and learning were composed and edited using social media and online publishing tools. Like *The Future of Learning, Learning through Digital Media* was reviewed with the CommentPress toolkit, and *Hacking the Academy* was aggregated largely from blog entries from nearly two hundred interested academics responding to the challenge to create "a book crowdsourced in a week." As editors Tom Scheinfeldt and Dan Cohen explained in their call for submissions at the start of the one-week collective authoring frenzy, "in keeping with the spirit of hacking, the book will itself be an exercise in reimagining the edited volume," so that any "blog post, video response, or other media created for the volume and tweeted (or tagged) with the hashtag #hackacad will be aggregated at hackingtheacademy.org."[43] Both books profess to being deeply concerned with changing power relations in the university and claim to emphasize concrete practices rather than abstract principles.

Scheinfeldt and Cohen's *Hacking the Academy* opens with a series of provocative questions that suggest that professors and other senior experts might be outsourced entirely: "Can an algorithm edit a journal? Can a library exist without books? Can students build and manage their own learning management platforms? Can a conference be held without a program? Can Twitter replace a scholarly society?"

> As recently as the mid-2000s, questions like these would have been unthinkable. But today serious scholars are asking whether the institutions of the academy as they have existed for decades, even centuries, aren't becoming obsolete. Every aspect of scholarly infrastructure is being questioned, and even more importantly, being hacked. Sympathetic scholars of traditionally disparate disciplines are cancelling their association memberships and building their own networks on Facebook and Twitter. Journals are being compiled automatically from self-published blog posts. Newly-minted Ph.D.'s are foregoing the tenure track for alternative academic careers that blur the lines between research, teaching, and service. Graduate students are looking beyond the categories of the traditional C.V. and building expansive professional identities and popular followings through social media. Educational technologists are "punking" established technology vendors by rolling their own open source infrastructure.[44]

Sections of *Hacking the Academy* are headed with titles—such as "Scholarly Societies and Conferences," "Academic Employment, Tenure, and Scholarly Identity," and "Departments and Disciplines"—that suggest that these institutional interests might be ready for the reforms recommended by the authors.

Although the table of contents of *Learning through Digital Media* reads like a how-to manual that is organized according to one's interest in particular tools, editor Trebor Scholz is adamant that instrumentalist-tool literacy approaches do little to respond to or effect cultural change.

> Learning with digital media isn't solely about using this or that software package or cloud computing service. The altered roles of the teacher and the student substantially change teaching itself. Learning with digital media isn't about giving our well-worn teaching practices a hip appearance; it is, more fundamentally, about exploring radically new approaches to instruction. The future of learning will not be determined by tools but by the re-organization of power relationships and institutional protocols.[45]

Scholz argues in *Learning through Digital Media* that such learning is characterized by an enthusiasm for cocreation and a synergy with informal peer-to-peer teaching that can "prepare learners for democratic citizenship," "community development," and critical engagement with the world.

As a professor, Scholz is cognizant that efforts to remake the university have a long history that has unfolded differently on multiple continents. He notes that his own faculty appointment is located at the New School, which was founded by critical theorists from the Frankfurt School who fled Nazi Germany to find a new intellectual home in the United States, where they treated their new institution as an experiment. Similarly, free-software activist and artist Florian Cramer has attempted to raise awareness in the United States about pedagogical experiments conducted by European artists who have undertaken rethinking the academy and the meaning of free and open education.[46] As an example, he cites the Free International University for Creativity and Interdisciplinary Research, which was founded by Joseph Beuys in 1972 as a form of "social sculpture."[47] Later iterations include the Situationist-inspired Copenhagen Free University, started by Henrietta Heise and Jakob Jakobsen in 2001, which operated with a spirit of satire, critique, and play before folding after posting an online statement of victory in 2007.[48]

Gary Hall's prescription for "The Free, Libre, Open University" emphasizes the importance of other kinds of experimental universities.[49] Hall is a faculty member at the Coventry School of Art and Design, a key member of the Open Humanities Press, an advocate for the Liquid Theory movement that promotes the circulation of media gifts, and the author of radical academic manifestos such as *Culture in Bits* and *Digitize This Book!* His work challenges the success story of Britain's Open University (OU), a pioneering institution in publicly funded distance education that is now the largest educational provider in the United Kingdom. Hall claims that initiatives like "The University of the Poor" or the "Really Open University" do more to counter the forces of neoliberalism and the hegemony of global capital than the OU, because these initiatives, which use tactical media to further social critique, focus on deeper issues of justice and effecting structural change rather than on more superficial issues such as access to standard curricula or providing services to fulfill immediate needs.

On its website, the Really Open University (ROU) explains that it is located in the city of Leeds and characterizes itself as "non-hierarchical and open to anyone who wishes to see an end to the commodification of knowledge and the creation of a free and empowering education system where creative and critical thought is fostered."[50] Like the ROU, the University of the Poor claims that universities have traditionally favored elites and preserved the status quo. The University of the Poor official website describes the university as being first conceived in 1999 and narrates a history in which it has "shared education tools with grassroots anti-poverty groups across the nation, has helped link media professionals, performing artists, social workers, members of the religious community and labor organizers with the movement to end poverty, and has facilitated the exchange of knowledge between poor people and their allies across the globe."[51]

In thinking about the "ownership of knowledge and reproduction of the labor force," Hall emphasizes the value of radicalized sites that draw attention to what he calls "cognitive capitalism," such as Edu-factory, which looks at the central role that universities play in capital economy. On its website, Edu-factory identifies itself as a "transnational collective engaged in the transformations of the global university and conflicts in knowledge production."[52] Many such organizations have forged ties to the Occupy movement, the Pirate Party, and other transnational groups devoted to fighting against globalization and plutocracy. Other projects Hall references include the Workers' Punk University, Imaginary Border Academy, and the University of Openess.

Hall also expands on the concept of "the unconditional university" imagined by French philosopher Jacques Derrida.[53] Hall observes that, because any type of university, no matter how open, is still tied to medieval structures of authority and to Enlightenment ideas of authorship and attribution, the "open university is actually very conditional" as it engages in publishing and disseminating work.[54] The university's moral philosophy concerning piracy is too culturally conservative for Hall, and he predicts that neither capitalism, communism, nor the gift economy will dominate in the institutions of the future that accommodate the practices of online culture. Hall cites Hardt and Negri's work on the proletariat in *Multitude*[55] to imagine serving a heterogeneous aggregation of migrants, workers, social movements, and nongovernmental organizations in the free, open, *libre* university. Hall's university would offer accredited degrees to students, and would thus differ profoundly from lifelong learning resource centers for autodidacts, such as The Public School, which proudly celebrates having "no curriculum," being "not accredited," issuing no degrees, and lacking any affiliation with the public school system.[56] Unlike utopian efforts by disenfranchised antiacademics, Hall would use the resources of existing institutions and harness the anger of the university-dependent middle class.

While *Learning through Digital Media, Hacking the Academy,* and Gary Hall's manifestos argue for substantive changes to the university's traditional association with print culture, particularly in the area of copyright, the pedagogical revolution championed by Douglas Thomas and John Seely Brown's *A New Culture of Learning* advocates an even more radical transformation. The authors frequently turn away from the university and print culture entirely to embrace informal learning cultures in which collectives of nonacademics work with shared, tacit knowledge, such as the guilds of massively multiplayer online role-playing games. Although Thomas and Brown insist that they "do not argue that classrooms are obsolete or that teaching no longer matters,"[57] they reject concerns about traditional literacy as outmoded trivialities. Citing the statistic that 63 percent of Americans aged eighteen to twenty-four can't find Iraq on a map of the Middle East, Thomas notes this task is easily performed when a computer is handy. Thus the authors redefine "where" questions as encompassing knowledge about how to tap into appropriate search aids and contextual clues rather than being limited to mastery of the absolute truths of geopolitical geography.[58]

However, the stories that Thomas and Brown tell about the wealth and freedom of collectives of remote learners only include enthusiastic and positive responses from participants. Interesting failures and the messiness of success are absent in their accounts of online learning. For example, the online community for the kids' computer programming language Scratch is described by Thomas and Brown as a place where "the single most important thing was to 'not be mean' in your comments" in promoting a "game" that "cultivates citizenship" rather than just teaches programming.[59]

However, MIT's Mitchel Resnick, the creator and chief administrator of the Scratch website, reports a much richer range of interactions between learners, including preadolescent male boundary testing, expressions of hurt feelings, and efforts to subvert totally justified forms of social policing.[60] Thomas and Brown also praise the microfinance site Kiva for its model of distributed charity and democratized wealth that supposedly provides a 98 percent repayment rate to investors;[61] yet they overlook the more complex picture drawn by more disinterested researchers demonstrating that Kiva boosters are using evidence selectively.[62] Wikipedia is held up by Thomas and Brown as "one of the best examples of the new culture of learning,"[63] but the problems that Wikipedia is having attracting new editors to contribute to the site, addressing gender bias, or developing a sustainable financial model are nowhere represented. And the possibility that one might choose *not* to participate or engage at all, the most common scenario in Internet traffic, may be analyzed in depth in Geert Lovink's *Zero Comments*,[64] but Thomas and Brown's netizens are always shown as eager and committed and never shown as cynical or detached.

Abolitionism Outside the Academy

A number of books by journalists who write about technology and education have been garnering attention from a public concerned about rising tuition costs, rapidly changing career paths, and quickly evolving trends in computerized delivery systems. If books by academics sometimes emphasize history and the past of education, books by instructional technology reporters are more likely to focus on the future. In these accounts, university administrators are sometimes depicted as villains, cold fish, or buffoons; imaginative and resourceful entrepreneurs, who appear as heroic protagonists offering salvation to students otherwise trapped in the status quo, are the main sources for information. For example, Jeff Selingo's *College (Un)bound: The Future of Higher Education and What It Means for Students*[65] has been described by one reviewer as "a love letter to ed-tech companies" by a writer "caught up in the celebration of genius expressed through technology, while the genius of 'ordinary' classroom teachers is minimized."[66]

Anya Kamenetz's *DIY U: Edupunks, Edupreneurs, and the Coming Transformation in Higher Education* declares that higher education is either unnecessary or unaffordable for most Americans and that computer-based learning offers more economical, customizable, and measurable types of instruction than what is currently available at most brick-and-mortar institutions.[67] As a journalist, Kamenetz's primary identity is as a consumer advocate and not an educator. Her first book, *Generation Debt*, focused on the high costs of a college education, and her second book addresses what she sees as its limited benefit. In sketching out a coming crisis over easy money for all applicants, like the one that took place during the mortgage meltdown, she argues that the tuition bubble eventually has to burst.

In her extended diatribe against the traditional status quo in *DIY U*, Kamenetz makes her case with a series of colorful analogies. Higher education is compared to the health care system,[68] the weight-loss business,[69] the airline industry,[70] exclusive chain stores peddling status brands,[71] and a world religion.[72] Technology provides speed skates, rather than crutches, for those who wish to learn,[73] and fast-paced new venues for learning are more akin water polo than the patrician game played with polo ponies.[74]

However, in her advocacy for "deschooling" or "unschooling" based on learning through positive and supportive peer feedback, Kamenetz seems to imagine a world without trolls and frauds on the Internet—one where the privacy of students and the credibility of authority figures in the classroom is no longer needed. She too often refuses to distinguish between homegrown DIY efforts, such as the Khan Academy, and corporate, for-profit predatory business models in a text that includes sometimes fulsome praise for universities following the Kaplan model of online education. Despite the work of teams of investigative journalists who have exposed many of the problems and potentials for corruption inherent in the "College Inc." high-pressure marketing model of sales-driven, consumer-oriented education, Kamenetz does little to acknowledge counterarguments and minimizes the objections of potential critics.[75]

Journalist Jeff Jarvis explicitly uses Google as an exemplary model to argue for radically remaking university structures in *What Would Google Do?* Jarvis uses fiery oratory to accuse colleges of being relics of the Industrial Age and its assembly line paradigms devoted to "stamping out everything the same: students as widgets, all the same."[76] Even as a faculty member running a center for journalism at the City University of New York (CUNY), Jarvis claims that both education and the media only "validate" and "repeat."[77] A self-styled iconoclast speaking as a guru for high-tech personalization and customization online, Jarvis claims to distrust any "right answers." Even at the TEDxNYED conference, where Jarvis was among many like-minded radical abolitionists, he expressed strong dissatisfaction with the inherent conservatism of the movement, calling "bullshit" on the lecturer/audience format and the artifices of the TED-style event.[78]

Although Jarvis claims to be grounded in practice rather than theory in *What Would Google Do?*, he ignores problems, inconsistencies, and design flaws that any educational pragmatist would immediately see. He hypes Google Scholar despite its beta characteristics. He believes that his educational utopia should not only offer students unlimited choice of instructor but also give instructors unconstrained power to select their own students. Jarvis lauds the successes of famous dropouts from elite colleges who have started high tech companies, but he refuses to acknowledge the value of those colleges as springboards for networking that probably made it possible for the future dropouts to start banking social capital from their first day on campus. He recommends online portfolios unequivocally, even though internationally renowned experts on such portfolios, such as Kathleen Blake Yancey, warn about potential problems in implementation that may occur unless students receive clear instructions, including guidelines about producing materials in new genres, providing rhetorical context for specific professional audiences, and directing thoughtful reflection that provides a meaningful interpretation of both the products and the processes of one's education.[79]

Jarvis insists that moving "customers up the design chain," as tech companies do, would inevitably offer students more choices. Yet such consumer-driven education could quickly sacrifice valuable academic activities such as contemplating, reflecting, questioning, doubting, and exploring unpleasant truths. Furthermore, satisfying individual desires, which could be manufactured to create a more profitable supply-and-demand market, could take precedence in education over serving the public interest. To be fair, Jarvis does acknowledge that "education should not always belong to the student,"[80] and he recognizes the importance of external frames of reference and neutral standards for truth. But he claims that the Internet "treasures facts and data" and thus serves self-regulating functions that make academia unneeded.[81]

The Googlization of Education

In debating Kamenetz and Jarvis, Siva Vaidhyanathan, the author of *The Googlization of Everything*,[82] has expressed strong doubts about entrusting education to the predilections of commercial companies. He specifically warns against numerical optimization strategies driven by the opaque black box of private companies and their proprietary algorithms. He has also mourned the public failure represented by the abdication of responsibility by policy makers that has created a large vacuum that social media and distance learning companies can now fill with narrowcasting practices in which human endeavor becomes devoted to pleasing niche audiences.

Using Ray Oldenburg's idea of a "third place"—which is neither home nor work—from his classic work on community building, *The Great Good Place*,[83] Vaidhyanathan claims that the classroom actually functions as a "fifth space" that is neither private space, public space, commercial space, nor space controlled by the state. Even when the campus is a state-supported organization funded primarily by taxpayer dollars, it functions as a "sacred space" that reflects the university's traditional origins in ecclesiastical institutions. [84]

In a keynote speech at the CUNY Graduate Center for the 2010 conference on "The Digital University," Vaidhyanathan characterizes campuses as remarkably "willing to experiment."[85] He claims that the impact of e-mail and course websites has been long undervalued as a source of "necessary intimacy" with students that levels predigital social hierarchies and spurs critical discussion among faculty in an atmosphere of greater transparency and access. He argues that austerity and commercialization are simultaneously driving the pursuit of ambitious digital projects rather than an interest in true experimentation and learning about learning. He defends tradition and ritual as ways of preserving diversity in the institution, and argues that too large a public profile—thanks to the exposure of the Internet—inhibits faculty performing the crucial roles of intellectual trickster and devil's advocate. He asserts that teaching is not about "delivering information," as Jeff Jarvis claims, and questions the YouTube/iTunes paradigm of consumption.

In making the case for offline learning, Vaidhyanathan notes that particular teachers' classroom tactics could easily be misconstrued and that "students should not have their words trapped forever," since a learning population should "try on ideas for size." Furthermore, Vaidhyanathan asserts that "capturing" and "publicizing" every moment in the pursuit of fame only perpetuates the existing focus on elite institutions. He describes "overflow" at brick-and-mortar open-admissions institutions, such as community colleges, and calls for greater investment in face-to-face teaching within current mass-teaching, educational-democratization efforts.

When Vaidhyanathan faced off against Kamenetz at Baruch College in 2011, he rhapsodized that universities are "national treasures" representing a "public good" that draws students from all over the world and contributes innumerable assets through research initiatives.[86] He described students as "partners in discovery" not "customers," "our product," or "victims," as Kamenetz might argue. Kamenetz began by saying that she didn't care about the future of the university and declaring that she cared instead about learners who struggle with an educational system that she characterized as "unsustainable." Soon she was alluding to familiar sources arguing for radical change in the university: access evangelist David Wiley, promoters of the open courseware initiatives at MIT, and advocates for self-organized or informal learning more generally. She also gave a plug for the National Center for Academic Transformation, which features an image on its home page of the familiar diagonal-slash-within-circle icon of prohibition superimposed on a photograph of students listening attentively in a college lecture hall.[87]

Vaidhyanathan countered by rebutting Kamenetz's truism that the university had not really altered since the founding of the University of Bologna in 1088 by citing not only the availability of new "gizmos" in classrooms but also the prominence of new teaching policies and philosophies developed during a period of energetic experimentation inside the academy. In response to the moderator's questions about accreditation,[88] Kamenetz suggested a model following the recommendation systems of social network sites such as LinkedIn, rating systems on sites for consumer products and services, and online communities of programmers, such as GitHub, that encourage peer evaluation.

When Kamenetz suggested that Vaidhyanathan should go into business for himself and leave the university behind, following the model of the free enterprise start-ups, he scoffed at the comment and characterized such schemes as wasteful and inefficient. Perhaps the most emotional moment in his rhetorical performance involved India, the country of his father's birth. He depicted India as a "nightmare higher education system" composed of a few "technical institutes" and "fly-by-night," "low time-investment certificate programs" subject to scams in a market-dependent society without public infrastructure.

Much of their debate involved negotiating positions around their shared use of particular forms of figurative language. For example, the Kamenetz–Vaidhyanathan conversation frequently compared the classroom of the future to a Starbucks branch, with she using the analogy to argue for the importance of new kinds of public spheres[89] and he warning against impersonalized mass-market consumerism. Ironically, when Kamenetz refers to Starbucks in *DIY U*, she suggests that the coffee chain is an icon of the snob appeal and misplaced brand loyalty that gets associated with universities.[90] Kamenetz also challenged Vaidhyanathan's metaphor of the sacred space to argue for a "reformation" that could offer alternatives leading to enlightenment without any kind of institutionalization. He countered by observing that Luther himself was a scholar whose reflection was made possible by institutional protections. Of course, such rhetorical uses of church vs. anti-church dichotomies in debating the status of education are not particularly new. Many others have used Eric S. Raymond's 1999 book on open source software engineering, *The Cathedral and the Bazaar*,[91] to explain dueling paradigms in everything from K–12 schooling[92] to the most advanced forms of law school education[93] and to make the case for open, egalitarian bazaars of learning rather than closed, hierarchical cathedrals of education.

Vaidhyanathan has also publically debated antiacademic firebrand Jarvis. As Vaidhyanathan notes, universities created Google, but Google is much more poorly equipped to create universities. Jarvis may be confident that "all the world's digital knowledge is available at a search,"[94] but archivists who specialize in how the public record is imperfectly stored online point out that the maintenance of digital collections often depends on human curators and the availability of financial resources. Moreover, even if content is "live" on the Internet, there is a lot of dark matter on the Web that search engines can't or won't see. Jarvis constantly differentiates the supposedly unfettered freedom of the Internet from the constraints of the world of atoms, as though computer chips and transatlantic cables are unnecessary for a global communication network. He describes online learning as fluid and efficient—as though outages and errors never take place.

Who Speaks for the Students?

If there are many practitioners arguing that the university is in crisis and society's resources must be marshaled to save it, there are also many arguing that informal learning through digital media is flourishing and that the demise of outmoded institutions of higher learning consequently should be hastened. Many have pointed to the images of alienated lecture hall students holding up signs that express their disengagement with academic work depicted in Michael Wesch's *A Vision of Students Today*, which has received over four million views on YouTube, as evidence that the conventional college experience and its apparatus of lecturing, textbook reading, and testing is failing.

The two hundred enrolled students in Professor Wesch's 2007 Introduction to Cultural Anthropology class who participated in *A Vision of Students Today* present a spectacle of minimum energy expenditure for college credit that is difficult for pedagogues teaching undergraduates to ignore. As lilting electronica music plays in the background, stone-faced students silently hold up signs with grim statements and statistics asserting that their online experiences occupy far more of their attention than their mandated coursework could ever do. They complain, via the signs they hold up, about expensive textbooks that they never open and classmates who never bother to attend lectures that they consider trivial. On-screen titles tell us that these Kansas State students created the very film that we are watching through a process of writing based on a collective-intelligence model, which resulted in 367 individual edits that were made to a shared Google Doc for the project.

Although they seem to be active participants in creating a collective student-centered manifesto for the video, Wesch's students assert that comparatively little reading and writing is taking place in service of their coursework in other classes, especially when measured against their engagement with literacy practices associated with social media and other forms of computer-mediated communication. The following four statements that are written on signs and held up by students in the video demonstrate the dramatic difference in their productivity between their online and offline lives.

I will read 8 books this year 2300 web pages & 1281 FaceBook Profiles
I WILL WRITE 42 PAGES FOR CLASS THIS SEMESTER AND OVER 500 PAGES OF E-MAIL
I FACEBOOK THROUGH MOST OF MY CLASSES
I bring my laptop to class, but I'm not working on class stuff.[95]

Wesch's students insist that reading and writing in digital contexts occupies a different order of magnitude of their attention, and they imply that literacy practices oriented around traditional textbooks or academic papers can't possibly compete. Academic labor and intellectual inquiry is reduced to "class stuff," and Facebook becomes the most important part of the curriculum.

While most YouTube viewers have been focused on the message of high-tech anomie from these seemingly disaffected youth, the precise circumstances of the video's composition have largely escaped notice from the general public. As the instructor of record, it is obvious that Wesch frames his students' critical composition with his own professorial comments and epigraphs, so the students do not entirely create the film in which they star.

A few months after *A Vision of Students Today* appeared, University of Southern California writing instructor Mark Marino decided to extend aspects of this possible critique of the video produced by Wesch and his class to note the absence of students of color in the sea of faces supposedly representing "students today." He thus spurred a lively conversation about race, technology, and teaching. In a blog posting at *WRT: Writer Response Theory*, Marino claims that Wesch's *Vision* video not only continues the implicit argument about display technologies in an earlier Wesch video *Web 2.0: The Machine is Us/ing Us*, but also propagates pedagogical discourses in which—as Beth Coleman and Wendy Chun have asserted—race serves as a kind of technology, although users of cultural interfaces are often unconscious of what Marino characterizes as the "universalizing of whiteness."

Following up on his highly played and first Web 2.0 video, Wesch focuses this video on how today's students have changed with respect to their relationship to classroom technologies and technologies brought to the classroom (such as laptops, cell phones, and pens). In the video, Wesch uses superimposed quotations and other comments to make a point that seemed implicit in parts of "Us/ing," that technologies offer new possibilities but do not completely eclipse or erase previous technologies. ... In this case, he juxtaposes contemporary technologies with that pre-eminent display technology—the blackboard. Further, his students also become display media. Or rather, they show the ways in which they can use any surface, including the walls of the room, as sites of inscription, means of participation, directly contrasting the blank screens of their faces and their reports of less-than-full class participation.[96]

To make his counterargument visible to Wesch and to a broader academic public that would eventually include many readers of the blog for the *Chronicle of Higher Education*, Marino created a video remix of "A Vision of Students Today" called "(Re)Visions of Students Today,"[97] which he released on Martin Luther King Day as a reminder about the racializing of the digital divide. Marino used a simple digital effects system to make it appear as if the students are holding up very different assertions from the original messages on their signs—assertions such as "18% of my teachers are non-white" and "I will meet 8 people of color this year." After Wesch's line "Some have suggested that technology alone can save us," Marino asks "Who do we mean by 'us'?"

Wesch responded to Marino's video on his *Digital Ethnography* blog by pointing out that at least one student of color had been an active participant in the filmmaking process, despite the lack of diversity in the racial demographics of Kansas State.

On the day of filming, several students had ideas emerge right on the spot. Whenever they had an idea they would write it down on a piece of paper and hold it up for the camera.[98]

While Wesch's group was "reflecting on the size of the room and the anonymity this creates among students," one student held up a sign that said "I am more than just A FACE," which was followed by an African-American student holding up a different sign that read "There is more to me than just MY RACE." Wesch decided not to include this juxtaposition in the final film, a decision that he justified in the blog posting by saying, "It was a powerful moment, and the sign itself defies any simple reading."

Figure 2.3
Scenes not included in Kansas State University Professor Michael Wesch's *A Vision of Students Today*. Courtesy of Michael Wesch.

From Wesch's comments we are invited to consider two interesting aspects of the video that may be obscured by the way that it presents itself as user-generated content that was made collaboratively by systematically creating an electronic document that served as a source text. We learn that comments on the students' signs, rather than being products of the Google Doc we are shown in the film, may have been spontaneous day-of-filming performances that responded to the face-to-face dynamic of interacting with the camera as a piece of technology and the lecture hall as a stage.

We also learn that Wesch made conscious decisions about editing out some shots in the film for the final version of the public piece, which may imperil its previously assumed status as a wholly student-authored composition. As Marino has suggested, it may have been more interesting to give all two hundred students access to the database of clips shot that day and see how *they* might have edited it together differently from their professor.

Laptop Bans and Filtering Software

Some professors have pointed to Wesch's video to support their own calls for laptop bans in the classroom and a return to analog instruction. The most reactionary professors even call for disabling wireless access in the lecture hall, even if the technology is used by other faculty in instruction. After all, such reasoning goes, the students themselves declare in Wesch's video that they multitask and "Facebook" through classes. Why should faculty allow them to become even more disengaged by offering them unlimited access to digital distractions?

Faculty have also cited studies in scholarly journals that seem to indicate that the presence of computing devices inhibits learning and limits the retention of information from lectures. For example, law school professor Kevin Yamamoto published an article in the *Journal of Legal Education* describing the pro-laptop and antilaptop positions before citing evidence from his own course evaluations in an antilaptop classroom to bolster his assertion that others should follow his lead. Yamamoto argues that "verbatim typing" interferes with note-taking, even when students aren't "multitasking" by surfing the Web or checking status updates.[99] An article in *Computers & Education*, known for its generally pro-technology stance, looked at data compiled from "weekly surveys of attendance, laptop use, and aspects of the classroom environment" and concluded that "students who used laptops in class spent considerable time multitasking and that the laptop use posed a significant distraction to both users and fellow students."[100] The debate around the question of banning laptops in college lecture halls even spilled into the media mainstream thanks to coverage in the *New York Times*[101] and the *Washington Post*[102] that quoted exasperated professors and results from distraction researchers.

Some universities have gone so far as to install software that prevents college students from visiting the most popular social network sites, such as Facebook, much as most publicly funded labs in K–12 learning environments currently do. Companies, such as Netop, that produce monitoring and blocking software for computerized classroom spaces have expanded their marketing and sales to institutions with adult student populations. As Netop's marketing materials explain, the Internet "is full of potential for the classroom; it's also rife with distractions—which makes you wonder: how to restrict Internet access while still providing your students the powerful online tools they need." The answer that such companies suggest is to filter Internet content.

> The Site Filters tool lets you create a Resource List including all the websites you want your students to use, while putting the rest of the Internet off limits. For instance, it lets you show your class a particular YouTube video, without giving them free access to YouTube as a whole. ... Block Lists let your students use the Internet freely—just with a few exceptions. You decide which sites are inappropriate or distracting, and remove those only, while continuing to let your students tap into the creative resources of the Internet at large. Need to keep your students off Facebook while they're doing Internet research? Create a block list, add Facebook's URL and leave them to work without the distractions.[103]

The words "distracting" and "distraction" appear throughout Netop's materials, and the rhetoric of the company suggests that a technical solution can be applied to the problem of student attention. Thus the company's software could be seen as a persuasive technology that could counter the holding power of the branded products of other software companies.[104]

However, if technologically locked-down classrooms become the norm, indiscriminate use of this kind of software may have extremely negative consequences on instruction, research, and communication. Instruction is grounded in trust between students and teachers. Surveillance and constraint may violate an important perceived social contract of respect for students' abilities to self-direct—and might cause both students and teachers to lose face.[105] As Joseph Janangelo has argued, exerting "technopower" in academic environments creates an institutional environment closer to the panopticon of the prison than to the open forum of the university.[106] Additionally, many courses in various departments now teach about digital culture, and social network sites can be important objects of study. Decades after the personal computing revolution, definitions of scholarly communication are continually advancing to include many socially networked forms of online publishing—forms that expand how research, with rich citation of texts and media, can be shared with broader audiences. Furthermore, many university departments, research centers, and official hubs make regular use of commercial social network sites such as Facebook and YouTube.

It may not be surprising that social-computing behavior for minors is heavily restricted on K–12 campuses, given passage of the Children's

Internet Protection Act (CIPA), which requires schools receiving federal funds to certify that filtering software is used on their machines.[107] Although other types of restrictive federal legislation designed to protect minors, such as the Communications Decency Act (CDA) or the Deleting Online Predators Act (DOPA), have faced challenges from courts and constituents that have hindered full implementation, CIPA was upheld by the U.S. Supreme Court. State and local laws regulating Internet use by minors in educational settings often go even further, either by mandating that violators be disciplined or by focusing on the standard of "appropriateness" rather than only blocking directly harmful content. Henry Jenkins and I have argued that the use of filtering software in secondary education stymies the many positive learning practices that are evolving as students and teachers learn to use new digital tools; we have wired the classroom but hobbled the computer by blocking student access to the sites that they would tend to use most.[108]

The legal standard of *in loco parentis* that places university administrators in the role of caretakers for their younger charges can be difficult to negotiate, and it perpetuates the logic that college students should be treated like minor children. David Hoekema claims that the focus should be on fostering "moral maturity"; when campus authority figures attempt to regulate moral behavior by enforcing arbitrary rules, he argues, they are placed in the trivializing role of babysitters. To prove his point, he cites student-conduct manuals that cover such a wide range of how-to advice as to become meaningless.[109] If such documents attempt to explain the correct use of extension cords, condoms, and footnotes, as though these were all equally important for ethical participation in a scholarly community, Hoekema claims that more serious responsibilities—those which facilitate moral development as well as intellectual growth—are being shirked. If higher education moves to the appropriateness standard currently in use in many primary and secondary education settings and embraces a similar rhetoric around protection, there could be grave consequences for academic freedom as well.

Not surprisingly, current definitions of academic freedom already provide more rights for faculty than for students when it comes to free speech, and the leveling practices of digital culture seem to have relatively little effect on challenging divisions created by existing status barriers between teacher and pupil. As both consumers of Web-based content and producers of it, students generally have fewer recognized rights and protections than faculty. Administrators loath to meddle with academic freedom often say little about faculty Web pages, for example. Faculty content-creation on official institutional pages includes a heterogeneous collection of material that ranges wildly from the confessional to the satirical. On university Web pages, one can find testimonials about personal trainers with links to their gyrating bodies, narratives about dieting that show shirtless faculty, heartfelt warnings based on personal tragedy about never leaving infants unattended, step-by-step instructions for destroying marshmallow peeps with lab equipment, and obscene mock-scientific acronyms. Ironically, political blogging[110] and electronic civil disobedience[111] may create more serious problems for faculty perceived as radical subversives than would more obviously off-topic or off-identity material. This might seem strange given how academic freedom is often defined in terms of political tolerance and the open marketplace of ideas, but the Internet has also generated cadres of students policing political utterances by professors. Otherwise, administrators generally do little to keep the online conduct of faculty in check.

In contrast, college disciplinarians actively monitor student behavior on popular social network sites, and they often do so without any awareness of the complicated representational practices around identity politics, which have been described by researchers such as danah boyd and Amy Bruckman, that are commonly observed on such sites. Unfortunately, students' social media practices have generated a culture of reaction among many campus administrators, even when such user-generated content depicts fictitious or staged situations. For example, English graduate students at Iowa State found themselves in a standoff with their department chair after two satirical documentary-style videos called "The English TA Experience" were posted to YouTube.[112] In these parody videos, stock TA characters express stereotypical responses of apathy, arbitrariness, aggression, lust, and obsessive neurosis in connection with their obligations to provide academic labor by teaching undergraduates. When their university superiors requested that the videos be taken down, on the grounds of potential liability (since the clips seemed to celebrate a culture of pedagogical harassment and abuse), the grad students protested by creating a Facebook group that not only reposted the controversial videos but also reproduced an e-mail exchange about censorship and aesthetic merit in which another faculty member defended the graduate students' rights to free speech.[113]

Some students have sought legal remedies for intrusive Internet policing by campus administrators. Millersville University student Stacy Snyder sued the campus after being sanctioned for appearing in a costume party photograph labeled "Drunken Pirate" on MySpace. Even though the twenty-five-year-old Snyder was of age, concerned administrators wished to withhold her teaching credential on the grounds that her conduct was "unprofessional."[114] Similarly, the Foundation for Individual Rights in Education (FIRE) took out a full-page advertisement featuring T. Hayden Barnes of Valdosta State University in the 2009 edition of the influential *U.S. News and World Report*'s America's Best Colleges issue. Barnes was expelled after posting a photo collage on Facebook that protested a new parking garage, which raised the ire of the structure's proponents.[115] According to *Inside Higher Ed*, those at Valdosta who sought Barnes's expulsion tried, in a written reply to his FIRE lawsuit, to draw on what they mistakenly saw as incipient criminality in his Facebook profile:

> As additional evidence of the threat posed by Barnes, the document referred to a link he posted to his Facebook profile whose accompanying graphic read: "Shoot it. Upload it. Get famous. Project Spotlight is searching for the next big thing. Are you it?" It doesn't mention that Project Spotlight was an online digital video contest and that "shoot" in that context meant "record."[116]

At Ryerson University, first-year engineering student Chris Avenir also filed a 10 million dollar lawsuit against the university[117] two years after he was charged with 147 counts of academic misconduct after joining an online study group on Facebook in which students swapped chemistry answers.[118]

If the appropriateness standard is adopted, it would be difficult to define appropriateness universally and with adequate rigor. Each campus actor may be actively engaged in many types of expression of intellectual inquiry and debate—sometimes simultaneously on multiple computer windows. For example, Oklahoma State University currently defines "appropriate use" of computing and networking resources relatively narrowly within the categories of "instruction, independent study, authorized research, independent research, communications, and official work."[119] Yet the category of "communications" covers much ground, and work that is "independent" could vary considerably from specific course goals. It may be possible to ban laptops and lock down desktops, but as computing devices become even smaller, with smart phones that are impossible to ban, a return to the analog classrooms seems like a quixotic fantasy.

Time on Task

Another philosophy designed to control student behavior emphasizes digital busywork that would make it impossible to maintain both high grades and an active online social life. In-class students may be required to use clickers to show proficiency at digital pop quizzes, and out of class they may need to complete extensive online tutorials. Both kinds of activity are concerned with learning on the clock, although in one case a rapid response is rewarded and in the other case a much longer investment of online time is often tracked.

Now that so many administrators are making time on task a top priority, the turn toward course-management software is understandable. However, the trend toward monitoring and enforcing scheduled interactivity originated in the 1980s with higher education guru Arthur Chickering, who first popularized the "time on task" concept among his "seven principles of good practice in higher education" with the following famous formula: "time plus energy equals learning."[120] Many might argue that Chickering radically oversimplifies how learning happens by reducing it to a physics equation. What about the role of confidence in one's academic identity, engagement with particular research questions, aptitude for the subject matter based on one's prior educational history, or desire for a challenging curriculum or meaningful, world-changing instruction?

And what about the value of time *not* on task, when students can creatively synthesize coursework from multiple faculty members or reflect on past learning to ensure long-term retention of knowledge? Yet for administrators and faculty members concerned about the decreasing amount of time students spend on studying and schoolwork, particularly the decline marked since the landmark 1965 survey Project Talent,[121] the clock can provide at least one measure of student engagement.

Chickering argues that technology can serve as a "lever" to implement time on task more efficiently and that for "faculty members interested in classroom research, computers can record student participation and interaction and help document student time on task, especially as related to student performance."[122] He imagines a more efficient future for college students in which they can be coerced into more productive uses of their time by deploying instructional software.

> New technologies can dramatically improve time on task for students and faculty members. Some years ago a faculty member told one of us that he used technology to "steal students' beer time," attracting them to work on course projects instead of goofing off. Technology also can increase time on task by making studying more efficient. Teaching strategies that help students learn at home or work can save hours otherwise spent commuting to and from campus, finding parking places, and so on. Time efficiency also increases when interactions between teacher and students, and among students, fit busy work and home schedules. And students and faculty alike make better use of time when they can get access to important resources for learning without trudging to the library, flipping through card files, scanning microfilm and microfiche, and scrounging the reference room.[123]

According to Chickering, students' time should be reapportioned productively. Squandering opportunities on "beer time," looking for parking places, and now-antiquated forms of research in the library's physical spaces offends his sensibilities. For him, the architecture of the traditional campus presents only obstacles and detours; the virtual campus would benefit from the supposed seamless efficiency of tasks completed independent of geography. Chickering doesn't seem to have imagined a multitasking student in 1991, so time spent on one activity is also inevitably lost to another.

Chickering's economic model of drastically limited educational time and space continues to be influential, but I might argue that even beer time isn't necessarily wasted. After all, beer time can serve other cultural purposes: it can provide the social glue that connects undergraduates to peer groups,[124] and, like Harvard's famous house sherries, it might even lubricate discussion. Although I share the concerns of my fellow educators about binge drinking, I think it is important to be skeptical about using this metaphor, since beer time could include a late-night bull session about philosophy or great literature just as easily as it could be applied to a meaningless drunken frat party. Just as Jane Jacobs has argued that institutional puritanism denigrates the social contributions to cities made by bars,[125] we need to understand the whole ecosystem of student life intelligently—and this ecosystem will inevitably include at least some beer time.

Beer time is not the only use of time scheduled for removal on the calendar of the digital university: lecture time is also considered extraneous. If students and faculty can't maintain decorum, why subject them to the experience? According to the promoters of online instruction, the very genre of the traditional college lecture seems to be manifestly inefficient in the era of e-learning. Why have thousands of people teaching introductory chemistry every year at varying levels of academic rigor and professionalism in public speaking, when the single best performance could be recorded and shared with all? And how does a rhetorical performance of knowledge—in the form of a monologue—compare with more interactive modes of teaching that both accommodate learners at many levels and provide a continuous feedback loop of error-correction and benchmarks of progress?

Philanthropic organizations such as the Pew Foundation have been critical of the basic one-to-many format of the standard departmental subject matter talk for a number of years,[126] and pedagogues influenced by Brazilian educational theorist Paolo Freire resist the "banking" notion that students are merely passive containers who must be filled with information while they sit docilely in the lecture hall.[127] In recent years, educational reformers have embraced campus initiatives for project-based, problem-based, or inquiry-based instruction that encourages students to take part actively in the production of their own educational experiences independent of faculty performance. Before considering what a professorless higher education landscape might look like in the future, it is worth examining why professors still can command center stage in the online rhetoric of the present—and why administrators fear losing control in an era of viral video and Internet memes.

3 On Camera: The Baked Professor Makes His Debut

On September 6, 2006, Howard "John" Hall began what was supposed to be a typical lecture for his class at the University of Florida's Warrington College of Business. To students who were taking his course for credit via the school's distance learning program, the graying professor with wire-frame eyeglasses at first appeared as he normally would on camera, in a sport shirt against the blue draperies of the stage. Yet instead of providing a more conventional presentation about the history of management as a professional discipline, the supposed subject of the day, Hall began by looking at his audience conspiratorially and reciting the lines, "Listen my children, and you shall hear / of the midnight ride of Paul Revere." Then he continued seemingly to free associate for the next thirty-five minutes, during which time he recited the classic Boston rhyme about the Cabots and the Lodges, discussed the etymology of the word "khaki," argued with a female student about her travel plans, and talked about the origins of various profane words and obscene gestures. Even Hall's body language showed the degree to which he had abandoned all academic decorum. At one point he rolled on his back, laughing uproariously with his midsection exposed; at another he displayed his middle finger to the camera. About halfway through the first segment, Hall announced, "Life is not about business. Life is not about management."

As the cameras rolled, his performance was transformed into a digital video file that was destined for the Internet.

After the link to the digital file of Hall's disastrous online lecture was featured with derision on several popular blogs, and the video was reposted with titles such as "The Stoned Professor" or "The Baked Professor," the university moved into damage control to protect the reputation of its business program and distance learning itself from Hall's antics and their possible pharmacological origins. On the instructional technology Web log for the *Chronicle of Higher Education*, "The Wired Campus," administrators explained why they took what seemed to them to be their only course of action.

McCullough, senior associate dean at the college, said Mr. Hall had been relieved of his teaching duties, pending a review of his employment status. Mr. McCullough said he had received a half-dozen calls so far from "the curious public," adding that there are some "unfortunate things" going on with Mr. Hall. "This is a human problem, not an institutional problem," said Mr. McCullough. "This man has problems."[1]

The original video was soon pulled from the university's website, and the copies on online video file-sharing sites were removed by various hosting services after they were notified by the University of Florida that these videos appeared to be violating the university's copyright. As university spokesman Steve Orlando explained, "It wasn't really serving the primary purpose, which was an educational purpose."[2] The video continued to resurface, however, as a notable Internet meme throughout the year. It was recut as an abbreviated summary, remixed as a psychedelic version with digital effects, and parodied in other online videos on YouTube. People who had never taken any of his classes visited Hall's page on ratemyprofessors.com to give him glowing reviews long after he had been terminated.

In many ways, the memorialization of Hall's breakdown on the Internet represented a nightmare scenario for champions of the distance learning movement, many of whom have more recently appropriated the rhetoric of open-source advocacy and the idea that universities should become more transparent to the public and embrace decentralized distributed models.[3] In this way, the logic goes, a greater variety of charismatic pedagogues could offer up their teaching services via the Internet to life-long learners in the lay public without excessive administrative oversight or pressures toward conformity from peers that might hamper their efforts at outreach. Reorganization with digital efficiency in mind would also make the university more economically viable, advocates for distance learning argue.

However, even as Hall was achieving his ten minutes of eventual solo fame, it is important to remember that Hall wasn't entirely without minders in the room. Given the technical demands of this kind of multimedia distance learning course, a lecturing professor on stage may not be the only one managing the pedagogical show. Like many who teach in programs with online video, Hall was obviously accustomed to collaborating with his instructional technology personnel, who were apparently just off camera during his bizarre lecture. For example, in explaining what was supposedly a response to the threatened mutilation of English longbow soldiers, Hall appeals to his tech person, "Anthony," not to film the visual part of his explanation, since "I'm going to shoot them a finger." At another point, he asks "Belinda" to "zoom in on this." Credits at the end of the lecture also acknowledge the presence of camera operator "Jennifer Faries."

At times, there appeared to be method in Hall's madness in his lecture. For example, Hall occasionally refers to his prepared material for the class: notes about Assyrian commerce, quotations about the Code of Hammurabi, excerpts from early product liability law, and observations about what he called "management before the study of management" during both the Han Dynasty and the time of Machiavelli's *The Prince*.

A Professorial Protest?

What's most interesting to me about this story is the *second* hour of the lecture, when Hall resumes the class after the break. Although still somewhat giggly and inclined to digressions, Hall is considerably more lucid. He also relies heavily on PowerPoint slides, in-class video, detailed lecture notes, and other prepared materials. In this section of his lecture, which few bothered to watch on the Internet, Hall explicitly criticizes the structures of the educational system and its attempts to rationalize the complexities of learning in a lecture devoted to explaining principles of scientific management and the legacies of its practitioners and theorists. As he describes the work of the pioneers of scientific management, Hall defends the value of historical context and, by extension, the humanities in the academy.

At first, Hall is merely pointing out the university's obvious role as a Weberian-style bureaucracy with a "clear division of labor," a "hierarchy of positions," "formal selection," "formal rules and regulations," "impersonality," and the separation of management and ownership from work.[4] For example, when talking about the growing size of organizations, he says, "think about the University of Florida" and describes meeting someone on an airplane whose nephew was a stranger to him among the over 3,500 other faculty employed at his home institution. Later, Hall expands upon this example.

> University of Florida is a bureaucracy. I don't need to tell you this. How many levels do we have? We go from faculty to department chairs to associate deans to deans to the provost to the president. I've lost fingers here. I don't know. Six. Yeah. A bunch.[5]

Although he repeatedly assures his audience that "bureaucracy is not such a bad thing" and "there are advantages to a bureaucracy," he also acknowledges that bureaucracy might produce "a bad taste in our mouths" with its insistence on written records and the division of labor.

Hall argues that bureaucracies respond poorly to major structural or systemic changes and that people trapped by regulatory strictures might be "S.O.L." or "screwed" when "the environment changes." During this second part of the class, he draws students' attention to particular architectures of control that are present before them in the educational system:

> I'm going to pick on myself. What about schools? What about schools? You know we had the model of the little red schoolhouse. How did we develop that? Let's think about this. Hello, can you say scientific management? Huh? Straight rows. Everybody has their own book. Everybody has their own test. What's the nature of work? What's the nature of work today? Is that what work is like? No. It's not what work's like ... We don't work like that anymore. That's not the way we work. But it's the way we teach, and it's the way we test, and isn't that baloney. Huh? Yeah. Isn't that baloney?[6]

Hall comically imitates assembly line work as he speaks about this antiquated paradigm for how labor should operate, and he asserts that current teaching and testing methods are as outdated as the repetitive motions that he mimes.

Hall's criticism of the educational system sometimes sounds more like David Noble's blistering attack on the digital delivery systems of impersonal higher education in *Digital Diploma Mills*,[7] or Frank Donoghue's critique of corporatization in *The Last Professors*,[8] than it does the kind of personal moment of failure or breakdown that is so often popularized on YouTube. It is ironic that the University of Florida has been much more careful to pull down from popular video sharing sites the first hour of Hall's lecture, rather than the second, on the grounds of copyright protection, even though the second part of the lecture reproduces a substantial portion of *Clockworks*, a 1982 film about scientific management and the role of stopwatches and cameras in improving efficiency, regulating labor, and imposing the model of the machine that repeats actions perfectly and rationally so that the bodies of workers and the intellectual practices of managers similarly serve as tools of integrated industry.[9]

Hall uses the film to introduce what seems to be an idiosyncratic cast of self-deluded characters, including Frederick Winslow Taylor, the author of *The Principles of Scientific Management*, the classic work on industrial time and motion study,[10] and Frank Gilbreth, Sr. and Lillian Moller Gilbreth, who were later made famous by the film about hyperefficiency in family life, *Cheaper by the Dozen*. The *Clockworks* film describes the obsessive and self-disciplining Taylor creating "a harness with wooden points" when he was "troubled by dreams" as a boy. The film's narration also notes that he suffered an episode of what might have been hysterical blindness. Although Taylor's work with the laboring classes may have led to insights about factory production methods, the irony of how excessive rationality produces madness does not seem to be lost on the so-called "baked professor." Hall then mocks Frank Gilbreth as a "nutcase." Although Hall presents Lillian

Gilbreth as a feminist heroine, he also ridicules her neologism for hand movements—"thirbligs"—and notes that her research was first applied to building the prison at Alcatraz.

Hall's criticism of Taylorization in education can be extended specifically to the distance learning situation in which he himself teaches and the application of principles of scientific management to creating modular lessons that supposedly deliver information in the most efficient manner. As Noble argues in the chapter on "The Coming of the Online University" in *Digital Diploma Mills*, the academic-corporate consortium Educause has performed a "detailed study of what professors do, breaking the faculty job down in classic Taylorist fashion into discrete tasks, and determining what parts can be automated or outsourced."[11] Donoghue insists that supposed "reformers" seeking to change higher education differ little in their aims from those described in Taylor's 1911 *Principles of Scientific Management* or the volume on *Academic and Industrial Efficiency*, which was published by one of Taylor's contemporaries and disciples, Morris Llewellyn Cooke, who championed for-profit universities, the abolition of tenure, and the deskilling of the academic workforce.[12] Donoghue situates his own book in the context of real reform against corrupting capitalism and places himself in a legacy of those who protect higher education, such as Thorsten Veblen and Upton Sinclair. Noble and Donoghue are not alone in criticizing the Taylorization of education; they follow the lead of other academics, particularly those in science studies,[13] who question the claims of objective neutrality that scientific management presents.

For many students and teachers, the advent of corporate courseware-driven e-learning only intensifies what the German philosopher Jürgen Habermas calls in the second volume of *The Theory of Communicative Action* the colonization of "lifeworld" by "system" under conditions of modernity.[14] I would argue that vibrant and successful classrooms—much like functional online learning communities—serve as lifeworlds by enabling dynamic discursive interactions with students in localized situations with constantly shifting themes and situational horizons. Education suffers when units of value in academia are reduced to variations of Habermas's central examples of standardization that privilege exchange value over use value: monetary currency and votes. Both simplify complex systems of value and suggest that everything can be measured by a single number. I would argue that page views have become the units of value in the attention economy, just as money functions in the traditional market economy and votes determine power in conventional political institutions. As Alexandra Juhasz has pointed out, online commercial platforms too often validate user-generated content only by popularity contests driven by viewer statistics, rather than by more nuanced means of legitimation.[15] It is noteworthy that the mainstream media tend to publicize notable success stories about delivering course content through online video by emphasizing the number of viewers as the main metric of success.

The Perfect Lecture?

At almost 15 million views, perhaps the most-watched video starring an academic on YouTube is Randy Pausch's "Last Lecture," which records one of the talks in the "Journeys" lecture series at Carnegie Mellon University. At the outset of his speech, Pausch notes that the series itself used to be called "the last lecture" and was based on the premise "if you had one last lecture to give before you died, what would it be?"[16] He follows this introduction by showing a CAT scan of his internal organs and announcing that doctors had found at least ten tumors in his liver, which represented two-to-five months of remaining "good health" in the face of a terminal illness from pancreatic cancer. After completing his hour-plus PowerPoint presentation about three topics—"my childhood dreams," "enabling the dreams of others," and "how you can achieve your dreams or enable the dreams of others"—he received a standing ovation from the crowd. He later partially reprised the lecture on the *Oprah* television show and appeared as a surprise speaker at the 2008 Carnegie Mellon graduation ceremony.[17]

Although I would not wish to denigrate Pausch's extraordinary skills as a rhetorician or the importance of creating materials in university lecture halls that appeal to the general public to whom we are all accountable, I think it is worth asking if this is really an ideal representation of an effective college-level lecture, if it models the appropriate level of engagement with conventional campus practices of research, critical reading, scientific demonstration, or the analysis and representation of data, and if it showcases the ethical standards of meritocracy that such institutions prize. From a number of perspectives, it is difficult to say that the "Last Lecture" really fosters an interactive learning experience, because it presents a spectacle of autobiographical achievement that was later packaged with a string bow and commoditized as mass-market motivational reading.

After all, the main subject matter of the lecture is Pausch himself. For the entire first half-hour, Pausch does nothing but tell the story of his life by running through a kind of virtual to-do list of his lifetime goals.[18] During this time, he displays his virility by dropping and doing pushups like Jack Palance on the Oscar stage, boasts of his ability to win large carnival-style stuffed animals, highlights his appearance in a photograph with celebrity William Shatner, and shows over a dozen childhood pictures of himself. Toward the end of the presentation, Pausch expands his focus to include his immediate family. He provides tributes to his mother and father and leads the audience in singing "Happy Birthday to You" to his wife.[19]

In the real world, faculty members at the podium who indulge in this kind of egocentric behavior on a continuing basis would likely receive very poor evaluations from students wanting to understand the specifics of material that exists in a more general frame of reference defined by objective facts. In telling his life story, Pausch only superficially glosses over specific experiments, software designs, or publications that might be on the examinations for real students. As one of the creators of the computer programming language Alice, which is designed to teach the principles of writing code to primary and secondary school students, and a developer of numerous educational games, Pausch is an important figure in the contemporary instructional technology movement in this country. As an advocate for public-private-philanthropic partnerships and interdisciplinary game studies programs, he does challenge the paradigm of the siloed research university in meaningful ways. But he never makes the case for the need for academic knowledge in the "Last Lecture."

If the aim of a successful college lecture is to facilitate the transferability of specific, disciplinary knowledge and thus empower the hearer as a social actor by sharing expertise, Pausch's lecture would seem to have a very different aim. He presents a fable about individual success rather than an introduction to a collectively shared body of knowledge. To be fair, he isn't necessarily making a rhetorical faux pas within the specific context of the "Journeys" series, since other faculty members have used the last-lecture format as an opportunity to reflect on the narratives of their lives, but it may not be the kind of last lecture that every pedagogue would choose to give or every administrator to sanction. Pausch also presents himself as someone who frequently flouts the rules governing academic conduct. A close reader may be surprised at the lecture's prominent place on the Carnegie Mellon YouTube channel, given the fact that he explicitly expresses hostility toward university values and institutional procedures several times during the "Last Lecture." In fact, much of his actual narrative of his adult life involves scenes of confrontation with authority figures who remind him about the unpleasant requirements or the conventions dictating what is appropriate for his professorial role. For example, when Pausch insists on going with his students into a zero-gravity environment despite a NASA policy forbidding his presence, he recasts himself as a local media journalist in order to be included and resigns from his sanctioned function as their faculty advisor. Later in the "Last Lecture," he describes siding with a Disney manager in a "culture clash" involving his own dean at the University of Virginia, who he calls "our villain" and compares to the nefarious "Dean Wormer" in the movie *Animal House*, even though his corporate collaborator observes that academics "are in the business of telling people stuff and we're in the business of keeping secrets."[20] In recounting this episode, Pausch even uses the potentially offensive terms "pissed off" and "pissing match" to describe the conflict. Later in the talk, he puts on a prop vest with arrows poking out of the back to complain about how his "pioneering course" was unappreciated by colleagues and how he subsequently stepped down from Carnegie Mellon University's Entertainment Technology Center.

In the section of the lecture in which he dispenses advice based on lessons learned, Pausch even describes subverting university procedures for undergraduate and graduate admission:

> Never give up. I didn't get into Brown University. I was on the wait list. I called them up and they eventually decided that it was getting really annoying to have me call everyday so they let me in. At Carnegie Mellon I didn't get into graduate school. ... And I was a bit of an obnoxious little kid. I went into Andy's office and I dropped the rejection letter on his desk. And I said, I just want you to know what your letter of recommendation goes for at Carnegie Mellon. And before the letter had hit his desk, his hand was on the phone and he said, I will fix this.

To those who respect the deliberative processes of admissions committees, Pausch's preference for pursuing special treatment might offend sensibilities. Looking at the transcript of what he actually says in his time on stage, it's worth asking if he would have received the same adulation if he had merely posted the words on a faculty blog rather than performed them on video for an applauding crowd as a dying man passing on wisdom in his own embodied voice.

A lecture Pausch gave in 2007 about time management is also posted to YouTube. Given his largely antiauthoritarian tone, it is remarkable to see the reverence that Pausch shows in this lecture for the University of Virginia as an academic institution, which he praises not only for its "tradition and history and respect" but also for its "honor code."[21] Unlike the scientific management experts featured in Hall's lecture, who rely on objective evidence from motion studies, Pausch locates the source of his own expertise in "what to do with limited time" in his "own battle with pancreatic cancer." However, like Hall's regulators of personal and professional life, Pausch equates time and money, criticizes "time wasters," and promotes regimes of scheduling and training. Significantly, his original plan for his talk focused on working with faculty advisors efficiently, but the version he delivers stresses the relationships of family and office rather than of the academy. It is interesting to note that this video received much less fanfare and currently lists fewer than forty thousand views.

In any case, it is difficult to ignore Pausch's legacy in the growing canon of classics in the genre of the online lecture. It is also important to observe how the forces of participatory culture on the Internet break down traditional town-and-gown barriers, particularly when the discourse in question is so personal as a public reflection on mortality in an embodied performance posted on the Internet. After Pausch's death, tens of thousands of people who watched the "Last Lecture" wanted to take part in the drama of the Pausch family's celebration and mourning and pay homage to the charismatic professor that they had seen on their computer screens. As *Wired* writer Alexis Madrigal points out in "Mourning the Internet Famous: Randy Pausch's Distributed Funeral," "grief-stricken comments" ran "for pages after every obituary or blog post bearing his name" and were often personally addressed to the deceased or to specific members of his inner circle. As Madrigal explains:

> The mourning also mimics the way that people experience Pausch's powerful oration. You interacted with Randy through a little box embedded in a webpage. Your headphones piped his voice clear and strong into the center of your brain, almost as if some deep part of your own mind was delivering his nuggets of wisdom. He was talking to you alone, not the hundreds packed into a theater or your family gathered around the television. In response, then, it made sense to get personal and say, directly, "Thanks, Randy. We'll miss you."[22]

Of course, John Hall's lecture appeared through the same platform of personal computing in the same online video format, but his combination of intimacy, personal vulnerability, physical performance, and use of stagecraft generated ridicule rather than respect. Both Pausch and Hall recorded a first-person address to a live audience for posterity, but they delivered two very different performances. Both questioned basic institutional structures in higher education and brought their personal lives into the lecture hall—but Hall was humiliated, and Pausch was lionized.

Rethinking Digital Literacy

Madrigal argues that the deeply felt fan practices that propagate Pausch's message split "the difference between the small and generally private funerals of our friends and family and the public spectacles that marked the passings of Stalin, or Elvis, or Princess Di." Because of Pausch's larger social media presence, such reactions might be understandable in the blurred territory between public and private life that has inspired so much punditry about the Internet. Status updates recorded both his radiation sessions and his swimming with dolphins, and he was connected through his Carnegie Mellon network to other active faculty users of social media who spread such updates, as well as to early adopters of instructional technology who might be experimenting with professorial *ethos* in new ways. His wife Jai wrote a memoir about being a caregiver, experiencing loss, and pursuing her own dreams separate from her late husband; however, she has taken pains to preserve her privacy in her socially networked public presence. Posthumous Facebook profiles memorializing Pausch continue to post updates and photos, so his social media presence flourishes, much as his frequently forwarded video continues to garner views. In contrast, John Hall has largely disappeared as an active content-creator on the public web, although search engine results indicate that he might be teaching for the online distance learning giant the University of Phoenix.

What causes a particular professorial viral video to circulate? In *Zero Comments*, Geert Lovink has described political pressures toward producing content that is combative or polarizing to gain a segment of the scarce attention to be had in the blogosphere and other social media venues.[23] In considering how faculty-created content will be vetted outside conventional channels of peer review, it is important to keep in mind that professors who are Holocaust deniers or promoters of other conspiracy theories will have built-in audiences, even though these individuals may be outliers in their own professions. Stanley Fish has proposed surrendering the idealization of academic freedom as a core value in the university to make regulating such unorthodox opinions more feasible,[24] but many fear that this approach will only make stakeholders more eager to eliminate constructive forms of dissensus in the academy and further undercut the authority of the university with the public.

Obviously, Pausch's video seems to avoid the sweeping political, historical, or scientific assertions that lead other kinds of professorial videos to receive millions of views. For example, Albert Bartlett, professor emeritus in physics at the University of Colorado, Boulder, has earned over 5 million views on YouTube with a video hyperbolically retitled "The Most IMPORTANT Video You'll Ever See." (The video's original title was "Arithmetic, Population, and Energy," and versions with this more descriptive and less histrionic heading received a considerably smaller share of views.) Although Professor Bartlett gives his lecture in a small lecture hall with barely a dozen students listening to his slideware presentation, his simple thesis—"The

Greatest Shortcoming of the Human Race Is Our Inability to Understand the Exponential Function"—turned out to have broad appeal to many latter-day Malthusians concerned about planetary population growth, and the video garnered over 14 thousand "like" ratings from viewers.[25]

Viral videos of professors going public in ways that can't always be contained by the constraints of the university raise a number of interesting questions as campuses contemplate moving more lectures to online delivery formats. As institutions try to maintain control over the voices and likenesses of their professorial employees and treat lectures as work for hire, questions about ownership and copyright have become important; but a long tradition of academic authorship independent of sponsoring institutions makes those in faculty governance largely resistant to any appropriation of their intellectual work. The University of Florida, with the aid of the Digital Millennium Copyright Act, largely erased Hall's performance; Pausch's 2008 copyright on the book version of *The Last Lecture* continues to provide funds to his estate. The significance of intellectual property issues in other professorial viral videos may be less obvious.

There are also a number of questions that show both the need for more rhetorically oriented approaches and the inadequacy of the faculty training currently being provided in professional development programs. How will lecture videos that are entertaining be judged in comparison with those that are informational? How will those that use academic evidence be judged in comparison with those that contain stirring testimony and personal revelations from the faculty member's own life? How will faculty performances that include live interactions with students be judged in light of current privacy rules?

In other words, the college lecture is more than a content delivery system for information that can be later tested. It is a performance of a particular identity position (or a series of identity positions) from which students can learn valuable lessons as they consider their own participation in staged arguments and public debates and develop eventually into authority figures themselves. The lecture has a long tradition in the history of oratory; for successful delivery, online lectures require more context and knowledge of the norms of new media genres, particularly if its platforms and interfaces are part of the user experience. Not only may shooting, editing, and compositing techniques be significant, but information about numbers of views, comments, and framing media may matter as well.

Students need more rhetorical training to make sense of increasingly digital forms of information delivery, and they need more training to improve their performances as online actors and media-makers themselves. Faculty may have their foibles exposed by easy recording, copying, and dissemination technologies, but students are vulnerable too. Even the most privileged students can become entangled in spectacular fiascos involving unintentional forms of Internet fame. One of the most notorious cases concerned Yale senior Aleksey Vayner, who sent out an unintentionally humorous "video résumé" to a number of investment banks. One of these video files was subsequently posted on the Internet and became fodder for a number of remixes and parodies.

The video shows a montage of Vayner involved in athletic exploits—weight lifting, skiing, playing tennis, tango dancing, and martial arts—that were entirely unrelated to the corporate job title that he was seeking. In the video, college student Vayner seems to be unaware of the inappropriateness of his pitch to his intended audience of corporate human resource managers in the professional world. He lectures his potential superiors about business success and competitiveness, emphasizes his lack of collegiality by calling others "losers," and exhibits a classic case of "sore winner" syndrome. With footage that could be described as a cross between an Internet dating video, a pitch from a low-budget motivational speaker, and a YouTube hoax film, Vayner lacks credibility both as a job candidate and as a representative of his prestigious college. Unfortunately, he probably only made matters worse for himself when he turned litigious and threatened his critics with legal action for violation of both his privacy and his intellectual property rights. Then those who had ridiculed Vayner carefully reviewed his print résumé and revealed a number of seeming exaggerations, falsifications, and fabrications. They publicized Vayner's fraudulent claims on blogs and social network sites, and coverage of the Vayner case in the *IvyGate* blog soon turned him into a national story. Although the *New York Times* published a largely sympathetic item on Vayner,[26] the *New Yorker* pointed out that he had a longstanding pattern of dissimulation going back to his freshman year and was widely mistrusted within his own social circles.[27] Later Vayner began to appear on the conference circuit to explain what he called the "context" of the video in speaking engagements addressed to those studying Internet ephemera. Notably, Vayner appeared at the ROFLCon conference at MIT, where he tried to save face by focusing on themes of privacy and crisis. He also focused on the uplifting themes of lessons learned about the value of loved ones and choosing one's own path. He argued that the video was initially shot to memorialize an influential martial arts master and was appreciated by the athletes he trained with.[28] He also advised those in a similar situation to be prepared to have an "outlet you can control" on "Web 2.0," such as a blog, to talk back to unintended audiences. Vayner lamented that he had "no opportunity to communicate with anyone" when his video went viral. By the age of twenty-nine Vayner was dead, allegedly from a drug overdose.[29]

One can create an illustrative tale of two college students by comparing the Vayner story to the experiences of Georgetown senior James Kotecki. Like Vayner, Kotecki was a white male who was in his final year at an elite university known for its network of potential professional contacts, but their narratives diverge fundamentally when it comes to their experiences with online video fame. Kotecki began by making videos that addressed presidential candidates and critiqued the content on their YouTube channels. The first of these videos was shot in his Georgetown dorm room and was addressed to one-time Democratic presidential candidate Christopher Dodd. In the video, Kotecki uses one of what would become his trademark pencil puppets of the candidates, but there is also substantive analysis of how well Dodd seems to understand the rhetorical conventions of the online video genre that he is trying to create. Campaigns began paying attention to Kotecki's videos. Candidates like John Edwards made videos back to the Georgetown senior, and dark horse contenders Ron Paul and Mike Gravel even came to his dorm room to be interviewed on Kotecki's webcam.

By the time of the YouTube debates, in which questions to potential presidential party nominees were generated from submissions to the popular video-sharing service, Kotecki had been profiled by the *Washington Post*, the *Los Angeles Times*, National Public Radio, and CNN. From his modest start with his Emergency Cheese channel, which eventually evolved into his own domain at jameskotecki.com, he found employment as a full-time professional paid commentator, first on Politico.com and then on The Daily. Despite the attention of the mainstream media, however, Kotecki is still keenly aware of the need to maintain narrowcasting norms and sustain the attention of his niche audiences. When he has appeared as a guest speaker at my digital rhetoric and digital journalism courses, he always offers to answer e-mails from individual students or add them as Facebook friends.

In an effort to generate more Koteckis and fewer Vayners, some colleges are experimenting with electronic portfolios, multimedia coursework, online publication of research, and hybrid teaching and learning initiatives to connect informal everyday social media practices to the discourses associated with academic literacy and print culture. Certainly, the consequences of neglecting the rhetorical dimensions of life lived in higher education and assuming that anyone can be a YouTube celebrity of the right kind with no preparation can be disastrous. Those who lack digital literacy and competence in digital rhetoric could find themselves economically and professionally disadvantaged, deprived of social capital, alienated from critical social networks, and unable to collaborate or solve problems effectively.

4 From Reality TV to the Research University: Coursecasting and Pedagogical Drama

You're listening to a UC San Diego Podcast
You're listening to a UC San Diego Podcast [speeded up]
You're listening to a UC San Diego Podcast [played backward]
Um. Um. Um. Um. Um.
Okay. Good question.
W-w-we-we found the number of moles here of oxygen.
Okay. Good question. Does that relate to the O_2? [in chorus]
We found the number of moles here We found the number of moles here of oxygen
Does that relate to the O_2? [slowed down]
Um
It will be more obvious
It will be more obvious
It will be more obvious
It will be more obvious
And
Many puzzled looks
Many puzzled looks [played backward]
Puzzled looks
Almost 75 percent unsure
[laughter]
[laughter]
See you next time
This has been a UC San Diego Podcast

This remix of a Chemistry 6A podcast was submitted as homework for my class on "Digital Poetics" in 2011.[1] A number of undergraduate writers in that creative writing workshop for computational media, which I first taught in 2011, chose course podcasts that the university made available as raw material for their individual audio compositions. Another student used a podcast taken from one of my own large lectures for an introduction to media theory course; it was disquieting to hear my own professorial voice similarly remixed.

Unlike the linguistic refrains in the Baked Professor remixes (see chapter 3) or the Angry Professor remixes (see chapter 1), these remixes bring out the poetry of everyday explanatory language rather than the shock value of memorable faculty tag lines—such as "my bad side is as bad as my pleasant side is pleasant" or "I'm going to shoot them a finger"—which could repeat endlessly in transgressive anti-professor memes. In the case of the student's podcast remix of the chemistry lecture, we hear the professor trying to sound authoritative by asserting the "obvious" character of the results, but we also hear uncertain laughter. Feedback, from both the students' "many puzzled looks" and the tally of what seems to be clicker response system results ("almost 75 percent unsure"), shows that the deductive chain of reasoning presented by the lecturer is failing with this population of learners. In these fragments from the lecture hall, we see the struggles of a typical lecturing faculty member captured for posterity, but these are the everyday struggles of conveying difficult ideas rather than the dramatic eruptions or breakdowns that YouTube loves to disseminate.

Although universities throughout the country often make such podcasts available to encourage students to listen or view lectures online either as a supplement to or a substitute for live instruction, relatively little has been done to conceptualize these podcasts as a specific rhetorical genre that might be distinct from the conventions that govern a live lecture hall performance. Currently, most podcasts on university sites are really recorded lectures, a byproduct of the traditional university experience rather than an actual delivery tool for learning. The artifacts of the chemistry podcast that the student remixed are certainly those of a live performance and a spontaneous interaction with an audience rather than a stage-managed digital show based on sequential modular learning.

In contrast, this chapter looks specifically at the genre of the coursecast intended for online audiences, which is recorded with an address to those remote audiences in mind. It asks questions about how the genre of the academic coursecast understands its own technical mediation, its own situation of reception through computational media, and its own intimate audience with the learner through the privacy of the personal screen or headphones. As distance learning becomes a larger stage for pedagogical drama, it is important to have a sense of how its norms shape rhetorical conventions about the learning situation more generally.

Pedagogical Drama

As a pioneer in video podcasting, MIT physics professor Walter Lewin achieved fame on YouTube for his dramatic demonstrations and accessible lecturing style. As Lewin said in an interview with CBS news, he believed that the pedagogical focus shouldn't be "what you cover" but "what you uncover," so that "students see through the equations."[2] Lewin well understood the elements of popular online videos that show DIY projects or impressive stunts. One of his most popular videos shows how paint cans and trashcans filled with water can serve as batteries. A lecture on angular momentum includes a stunt with a wheel seeming to defy gravity. Often Lewin's own body and the bodies of his students are used as props in the theater of his physics experiments. The photo topping his 1999 Classical Mechanics course, which was one of the most-viewed curricula at MIT's Open Courseware site, shows Lewin "demonstrating his faith in the Conservation of Mechanical Energy" as a large weight on a pendulum seems to come perilously close to his throat.[3] A promo from MIT's OpenCourseWare Initiative on YouTube shows Lewin swinging back and forth across the stage of a lecture hall.[4] As an article in the *New York Times* points out, at one time Lewin had also been a top star on Apple iTunes: "He was No. 1 on the most downloaded list at iTunes U for a while, but that lineup constantly evolves. The stars this week included Hubert Dreyfus, a philosophy professor at the University of California, Berkeley, and Leonard Susskind, a professor of quantum mechanics at Stanford."[5]

Although iTunes U has billed itself as a place that primarily "delivers easy, 24/7 access to educational content from hundreds of top colleges, universities, and educationally focused organizations across the country,"[6] it also serves an important function for admissions officials at universities, allowing administrators to showcase course offerings for prospective students who rely on mobile devices and ubiquitous computing technologies. However, the assumption that such podcasts can serve as recruitment tools might already be outdated, given the fact that incoming college students may no longer consider Apple's pay-per-item model as "wildly popular"[7] as it once was; new social media platforms and file-sharing services may undermine the single-provider model that connects an iTunes playlist of music to the offerings of iTunes U. The copyright complexities of podcasting pose many obstacles for university administrators, and many faculty have questioned if iTunes U is really a sustainable model.[8] After all, iTunes U is part of Apple's business plan, which is devoted to the proprietary licensing of content for profit, but universities are generally committed to the fair use of intellectual property and are often seen as public resources themselves.

Alternatives for the free delivery of course lectures have proliferated in the years since the launch of iTunes U. Coursera, edX, and Udacity have been developed with the active participation of faculty at elite institutions such as Stanford and MIT. Sometimes these initiatives have the full participation of the administration as well, and courses may be emblazoned with the distinctive brands of particular colleges. Other new ventures, with names like Udemy, Skillshare, and Dabble, appeal to a large population of so-called lifelong learners with less obvious ties to specific institutions. Some may be archives for both homegrown and instructional content harvested from elsewhere on the Web, as is the case with the Khan Academy, which famously includes a popular series of lectures on mathematics in which the disembodied, deep-voiced lecturer Khan occasionally acknowledges unanticipated events that interrupt his screencasts, such as a burp while speaking or the arrival home of his wife.

In the Coursera videos for his lectures on human-computer interaction (HCI), Stanford's Scott Klemmer presents a radically different personal affect as a professor from Walter Lewin's. Unlike the MIT physics professor, who presents himself as a body in motion, Klemmer maintains his scholarly composure as a seated body at rest. Klemmer's immobility might be punctuated occasionally by symmetrical hand gestures that emphasize an important point in the slides accompanying his lecture, but the camera is on a locked-down tripod, and Klemmer is careful to keep himself within the frame. Nonetheless, he attracted thousands of students who watched his short, snackable lectures on interface design, even though the only time they saw him standing upright was when he came to the Stanford computer science building in the dark, after hours, to show his viewers the bad interface design on a particularly vexing exterior door.

Although his lectures eschew the embodied oratory of the podium or the acrobatics of the stage, Klemmer, despite his low-key delivery, does seem to provide the necessary pedagogical drama to his students. After all, a quiet aha moment can create pedagogical drama, just as a crescendo or an explosion can. Students may also benefit from experiencing the sense of intimacy that the sight of Klemmer close-up creates. Not only is the scene of learning staged as though the learner is engaged in a one-on-one office hour; Klemmer is a skilled user of props from everyday life that are best displayed at relatively close range. For example, he holds up a measuring cup to the camera to illustrate how users get feedback about system state, or he demonstrates distributed cognition with three bagels of varying sizes.

Students are always aware that the presence of the professor is mediated by particular technologies. Coursera provides the videos of Klemmer's lectures through a specific interface of computer menus that remain visible in the background as the professor speaks. Klemmer himself comments on the poor design choices made in creating this interface, some of which can be attributed to Coursera and some of which reflect the first-time mistakes of

Klemmer and his cohort of teaching assistants and instructional technologists. At one point, Klemmer remarks on how the visual icons on the Web-based menu are extremely information poor, much as the generic icons on his credit union's home page convey much less information than appropriate text would provide.

> While good icons can often improve the usability of a site, generic icons rarely help. Another example of generic icons are the web page icons that you see along the left-hand side of this page here, which is the web site for the HCI class online. The generic globe page doesn't tell me about the link, and so it doesn't add any information. Those pixels could be better put to use with a more specific icon or with additional words.

Klemmer notes that the navigation for an essential course webpage is less legible to potential users because different elements—such as "Course Information," "Syllabus and Calendar," "Design Briefs," and "Peer Grading"—are all indicated with the same sphere icon that suggests little more than does the most indefinite globe in representing the broadest possible category of the World Wide Web.

Although he turns his students' attention to weaknesses in the online materials created by the interface designers at Coursera, Klemmer is willing to share responsibility for failures in communication and to emphasize the paradigm of constructive criticism and iterative revision that is central to the philosophy of his course.

> So how can we improve the scent of links? One of the most effective strategies for improving the scent of your website is to lengthen the length of words. ... One example of this is that when the HCI course began we had a link that I think was called "grading policy." Few people realized that behind that link was where we explained the different tracks of the course. And for good reason the word "track" was nowhere in the link. After a week of confusion on the forum, we realized this and were able to redesign the link to be longer and more specific and include multiple different trigger words that might bring people to that page.

The new name of the link on the navigation for the course webpage is ultimately revised to read "Tracks, Grading, and Statements of Accomplishment" in order to signal students that they should click on this category to find answers to perplexing and common questions about how they might receive credit for a course in which the instructor never sees—much less grades—their assigned work.

The format of Klemmer's coursecasts takes advantage of both changes to delivery systems for online video and shifts in user behavior toward greater acceptance for interactive activities during periods of viewing. It is worth

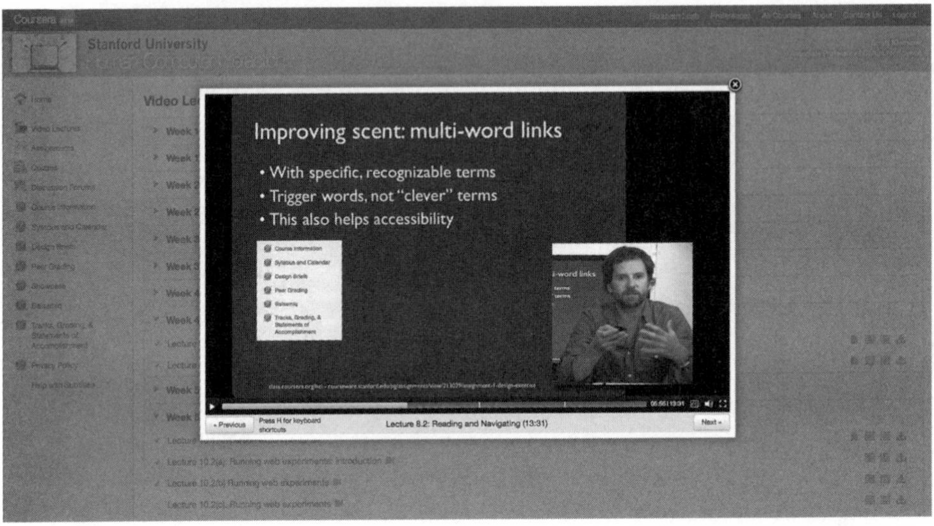

Figure 4.1
Stanford professor Scott Klemmer discussing the shortcomings of the Coursera site.

noting that Klemmer's students quickly embraced common techniques for crowdsourcing the task of subtitling or dubbing videos into other languages, so that those whose English skills might not be up to the professor's level have the opportunity to read the transcript of his words in their native tongues thanks to the availability of either machine-generated or user-generated content. As commercial online video sites like YouTube have integrated features like games and quizzes that are synched to particular points on the timeline of the playhead, users are becoming accustomed to clicking on embedded hyperlinks or adding annotations to video files. Videos like the popular 2009 "Barack Paper Scissors,"[9] which has received over a half-million views, are changing audience expectations about online spectatorship to include haptic as well as optical sensory engagement, and the norms of viewer conduct may be becoming active rather than passive in some Web-based video genres. Klemmer's Coursera videos have periodic multiple-choice quizzes embedded into the viewing experience so that students can gauge their comprehension of both major concepts and finer details during the lecture. However, these quizzes are experienced as interruptions in the pedagogical performance. Klemmer becomes abruptly silent, and the learner must reason alone without verbal help, hints, or prompts from the professor. In contrast, what would it mean to model other kinds of online exploration or trial and error as part of a continuous flow that facilitates multitasking, or at least tolerates some divided attention in each multimedia tableau, as is the case with certain online games or interactive puzzles in which video may be embedded?

The video coursecast is a difficult genre for faculty to learn to produce, and it is one for which there are few instructions, little mentoring, and almost no pedagogical theory. The instructional technologists who advise faculty often are not knowledgeable about methods of effective teaching from the practitioner's standpoint; in addition, they may have no attachment to either the specific research disciplines that shape students' performance expectations or the career paths of faculty looking to make progress in promotion and tenure. Unfortunately, instructional technology often imagines teaching as a delivery system rather than an interpersonal relationship.

The situation of bad professorial podcasts has become so dire that Associate Professor Stephen Ramsay of the University of Nebraska-Lincoln—with his university's support—is devoting time to creating a digital humanities tool designed to help faculty compose and edit pedagogical videos that, he argues, would be far superior to the impersonal, long-winded, and noninteractive recorded lectures that current distance-learning venues generally provide—even in the best situations (e.g., the video is being used in a hybrid context rather than as remote instruction). Ramsay has become known for promoting "flipped" classrooms in which the professor uses classroom time for mentoring hands-on practice; time that was once dedicated to homework can be spent reviewing the standardized presentations that generally take up valuable class time. However, Ramsay notes that the videos in flipped instruction are "seldom well crafted," which "is to be expected" because producing "high-quality narrative film—educational or otherwise" would require that "teachers work in an unfamiliar, and often highly technical medium."

> Teachers typically either film themselves giving a lecture, or else create a live voice-over of a screencast. The former method, barring the ability to retake, offers no better chance of reaching a teacher's ideal than a live lecture; the latter asks the teacher to give their best explanation while fumbling around with a computer desktop. The results, in either case, are usually very poor indeed. A professionally-produced narrative film, of course, would not use either of these methods. The instructor would record a voice-over, and then use video editing software to construct a visual narrative that is timed with the audio.[10]

In his application for leave, Ramsay asserts the primacy of a carefully drafted script in narrative film and foregrounds the technical ability to work with audio and video tracks separately to create a compelling and coherent experience for the viewer.

Ramsay argues that the popular animated-video creation website Xtranormal provides a good model from which teachers can borrow. This is because it can easily generate, from a written script, video content that is spoken by its cartoon characters. Xtranormal has already attracted some attention from the academy through social media dissemination by instructors. A number of its most popular cartoon videos present humorous send-ups of supposedly common teacher interactions with clueless students, such as "I Need an A"[11] and "One Professor's Fantasy."[12] There are also mocking videos of graduate students and would-be graduate students with unrealistic expectations about academia and the job market for English professors, such as "So You Want to Get a PhD in the Humanities."[13] Composition instructors are already using the site to create pedagogical videos for their classes, although the sight of instructors clad in Star Trek uniforms delivering advice about effective rhetoric may be disconcerting to students.

Ramsay notes that generating the kind of written text that drives Xtranormal is much easier for faculty than directing themselves in visually appealing videos—for which they would have be knowledgeable about everything from camera angles to motion graphics—would be.

> My idea involves uniting the written lecture/voice-over script with a highly simplified, domain-specific scripting language for generating text animations, video sequences, and displayed images. The narrative portion of the script could be recorded as audio, but then that audio file could be keyed to events described and programmed into the script itself. The video would then be automatically generated from the file. Such a system would allow an instructor to create a video for the flipped classroom using nothing more than a word processor and basic audio equipment.

Ramsay's idea for a push-button video production workflow is rapidly becoming a reality in a number of fields in which video can be generated automatically. In fact, experiments to create news broadcasts or children's cartoons algorithmically have been successful enough to be perceived of as a threat to the creative classes in infotainment or edutainment fields.

Coursera professors such as Stanford's Klemmer definitely do not need the kind of plug-and-play program that Ramsay envisions, because they have access to legions of instructional technologists seeking to create a polished product. However, finding a viable business strategy to support high-end production values and robust instructional databases for delivering what is ostensibly a no-revenue enterprise has posed many financial challenges to offering sustainable access to free college instruction. While edX charged campuses up to $250,000 per course in early 2013 for instructional technology services for design, recording, editing, and hosting, Coursera gained university clients by promising to do course setup at no charge to the content provider. Nevertheless, Coursera now appears to be marketing itself as a potential multimedia-enhanced course management system to vie with corporate giant Blackboard.[14] Such economic choices could shape a more standardized genre of coursecast for creating easy-to-replicate modular content, much as electronic gradebooks are designed to project neutrality and consistency to their clients.

However, as Klemmer suggests, courseware companies may sometimes lack basic expertise in user interaction design or, as I argue, in the rhetorical situation of dynamic pedagogy in a realm of shifting classroom interactions that stimulate discussion and debate. To better understand how pedagogical drama can function in online environments, I will point to two seemingly unrelated, powerful forms of stage direction in the present theater of learning: (1) the production of reality television shows devoted to developing professional expertise, and (2) the delivery of popular conference talks about epiphanies experienced by university researchers. After all, the rhetorical conventions of both broadcast television and traditional oratory still shape many newer and more snackable Internet genres, since content creators often attempt to appeal to older audience expectations, or at least allude to them as cultural references.[15] What may be less apparent is what faculty, in appropriating some of these same mediagenic directorial techniques, can actually learn about effective means of educating. Currently, coursecasts tend to feature talking heads that contribute little visual information and convey little affect to the viewer; this is so that students can gauge what is most important about the coursecast (the information being imparted) and internalize that information in long-term memory. Many students use only the audio portions of such lectures and may open other computer windows as the lecturer drones on.

Getting an Earful

The word "podcast" itself is still a relatively recent neologism, and one that privileges the Apple corporation over other platform providers in ways that don't always reflect patterns of media consumption on devices manufactured by other brands. Although much of this section of this book is devoted to rhetorical analysis of video coursecasts, audio can be a key channel for communication between faculty and students, as the story about the audio podcast told at the beginning of this chapter illustrates. Although audio podcasts have yet to become one of the electronic practices of everyday life for all Americans, asynchronous media consumption of sound files has become common enough that some users may not expect a live broadcast, much less a live performance, when it comes to hearing the voice of a lecturing professor. Recent patterns of media consumption have shown that online community building and social media connection may have enhanced existing recommendation systems for audio, and thus the audience share for podcasts continues to grow. A November 2006 Pew Internet & American Life poll indicated that only 12 percent of Internet users had ever downloaded a podcast and that a mere 1 percent said that they did so on a typical day.[16] By 2008, the number of podcast downloaders had jumped to 19 percent.[17]

As Richard Leppert contends in his essay "The Social Discipline of Listening," listening is not "a biological phenomenon"; rather, it is a "historico-sociocultural one," which—in the case of the expectations of nineteenth century bourgeois audiences for virtuoso musical performances—is framed by class and gender dynamics, by the economics of desire as well as those of consumer capitalism, and by the configuration of norms about public and private identity.[18] In a similarly historical vein, Lisa Gitelman argues that when media publics engaged with new auditory technologies—specifically, when the phonograph or the telephone was first marketed—there appeared to be a need for a prolonged educative period, if not a total rethinking of consumer use.[19] Mimi Ito's research in contemporary urban Japan likewise indicates that mobile electronic practices are social rather than individual in nature and that terms such as "personal," "portable," and "pedestrian" point to the role of a what she calls a "technological imaginary."[20]

In other words, cultural rules around group identities need to emerge in order to make particular consumer preferences for software applications on mobile devices function as standard operating procedures, which have evolved to the point of having sufficient associated systems of collective manners. In contrast to work being done about contemporary digital audio, an earlier iteration of academic criticism about the Sony Walkman focused on its supposed asocial and isolating character of cultural disengagement. For example, philosopher Michael Heim has described the "trance" associated with this technological device that brackets off the listener's subjectivity in an immersive "sonic bubble" or "electronic envelope" that "modifies the habitat" because "the ear takes in all directions at once."[12]

Highly successful early podcasts, such as "Tim's Podcast" for the reality television show *Project Runway*, would seem to contradict the applicability of the isolationist view to new forms of digital audio, since well-ranked podcasts depend on the existence of a complex social network of online and offline subcultures, fan communities, bloggers, or groups committed to a DIY ethos in various forms. There may be implied norms of individualism created by the situation of physical intimacy and phenomenological immersion for a solitary listener who experiences the digital files of the podcast through his or her personal computer or mobile device, but because this form of receptive literacy often involves the tropes of gossip and collective intelligence, that listener is never assumed to be the final link in the informational chain. Gossip—like replicable and alienable digital artifacts—implies the existence of secondary audiences, to whom seemingly subversive tidbits will continue to circulate. As Patricia Meyer Spacks has argued,[22] however, gossip generates and regulates discourse and thus serves a key normative function in public culture, and it can even function in the backchannel of pedagogical settings, where the public/private divide may be tricky.

Reality TV shows devoted to showcasing competition around the development of professional expertise can also show how teaching and learning take place collaboratively in the training of singers, dancers, chefs, decorators, and designers who contend for the top prize of assured publicity and financial backing in launching their public careers. Of course, it could be argued that such shows dramatize the worst kind of pedagogy, since the judges are usually passive spectators who provide little modeling or process-based instruction and who focus on error-finding activities that have been discredited by educational researchers.[23] Students on these shows tend to do little constructive peer mentoring, since they see themselves as part of a zero-sum game driven only by the pursuit of individual advantage. The politics of labor presented these shows—in which work is extracted without compensation and reputation serves as the main currency—are never seriously examined.[24] Yet we do see a developmental approach to the cultivation of specialized knowledge and skills in which imitation, iteration, improvisation, and invention all play a role, and innate genius or talent is deemphasized. Not all reality shows highlight such character evolution, since shows about romance or captive-living situations tend to emphasize expedient alliances between stock personalities rather than personal growth under the tutelage of others. And shows that model extremely hierarchical work relationships, defend the status quo, and promulgate the equation of corporate capitalism with leadership, such as *The Apprentice* or *Undercover Boss*, might also be problematic as paradigms for the higher education experience. As University of California, Riverside professor Toby Miller points out, the ridiculous spectacle of the chancellor of the university impersonating an athletic coach, library assistant, science adjunct, and other subservient roles on a 2011 episode of *Undercover Boss* merely reinforces existing narratives of paternalism, surveillance, and sentimentality and exalts the role of management, rather than learning in the institution.[25]

In online video, as in the case of its broadcast TV counterpart, the "reality" of campus drama is sometimes staged, as it is in the case of "angry professor" vernacular videos discussed in the first chapter of this book. Angry student videos also may be scripted rather than captured as spontaneous actions. For example, ABC and Fox News both carried stories about the fakery of the undergraduate couple at the center of the "UNC Pit Break-Up," who had hoped to capitalize on the hundreds of thousands of views earned by their video, which seemed to show a boyfriend dumping his girlfriend in front of a large campus audience of student spectators to tunes supplied by one of the college's female a cappella singing groups.

When it comes to representing the explicit values of the university, in many ways it could be argued that *Project Runway*'s Tim Gunn has served as a professorial role model in the world of reality TV by teaching to the crossover Internet audience. Throughout the first three seasons of the show, in his Web persona, Gunn asserted his identity as a former associate dean of the Parsons School of Design and the present chair of its Department of Fashion. Although Gunn could be read as a subversive figure, known for his hyperbolic comparisons and for his private counterdiscourses about race, class, sexuality, body type, and intellectual property, his blog and podcast validated a specific pedagogical orientation, one in which the reader or listener was encouraged to form allegiances to institutions in higher education. As such, Gunn represented certain scholarly norms, legitimated disciplinary practices in terms of academia rather than the stage or street, and included and excluded show members in key rites of professional initiation. He often expressed admiration for other "teachers" on the show, including contestant Nick who taught at a fashion college, and discussed degree-attaining students in admiring terms. He was concerned about protecting the physical plant of the Parsons space at The New School and preserving it from potential vandalism. In diplomatically negotiating conflicts between contestants, he frequently referred to his role as a mediator in the school.

In contrast, "Stand In" at MTVU presents a much more haphazard picture of teaching practices, which are captured by video when surprise celebrity guests come in to "sub" for a professor's class. Rarely, however, do these MTVU celebrities attempt to do the actual labor of their professorial hosts by arriving with a structured lesson plan. For example, Bill Gates didn't really "teach" the Introduction to Computer Programming class at the University of Wisconsin; instead, he participated in a light Q&A that included chatting about his favorite video games. George Clinton at the Berklee

College of Music also came without a prepared talk, although he participated in a jam session in which improvisation was perfectly appropriate. Madonna's appearance at Hunter College, in connection with the screening of a concert tour documentary, also depended on a Q&A format, although the singer knew to ask specific students questions. There were, however, some exceptions. In front of a Columbia class, Natalie Portman tried to engage students and cover concrete subject matter relevant to the assigned reading and the syllabus; for a class on art in society at Temple University, the professor praised the discussion fostered by rocker Marilyn Manson.

As one of the more successful early mainstream media podcasts, "Tim's Podcast" exploited an existing fan culture around *Project Runway*. Although designed to go with particular video content about novice fashion designers (similar to director's commentary on a DVD), "Tim's Podcast" also could function as a kind of stand-alone series, independent of the particular television episodes. Given that *Project Runway* is about fashion design and, therefore, would seem to be dependent on conveying visual information about the color, pattern, shape, and proportion of successfully executed clothing, it may seem strange for fans to have devoted attention to the audio-only podcast, but the information on the auditory channel was compelling enough to attract a sizable audience.

Obviously, part of the appeal of the podcast has to do with Gunn's verbal dexterity and virtuoso performances with language. Even in the video broadcast, on the reunion episode of season 2, Tim Gunn's over-the-top use of elaborate figures of speech and striking comparisons merited an entire segment of recaps on the show, where his references ranged from confectionary food items to amateur-themed entertainment by historical reenactors. In episode 2 of season 3, his associative chain followed a particularly circuitous route through different systems of signification to describe a gown being made for the reigning Miss USA for a red carpet event.

> I saw this big top, big giant top. I remembered saying that it, and the fabric, the look of it, the ruching … that it looked like it had been carved out of a log … as I think of it now, it was like a giant piece of fudge that they had formed. A confectionery it was not, leaden it was, like my mother's fudge (forgive me mother) … She is the second shortest Miss USA in history … this is not a tall woman, so she doesn't want to look like she just came out of the forest primeval as not a water sprite, but as a Yule log.

Later in the same episode, he describes this dress as both "a lumpy sack" and "hot mess." He characterizes another attempt as looking like a "column," a "Beam me up Scottie" design suitable for "Judy Jetson's birthday party." To capture these verbal performances around the mixing of metaphors and render them legible to search engines and to the activities of textual analysis that characterize the discursive practices of many text-oriented Internet communities, a volunteer cadre of fans transcribed many of the podcasts and posted them to the fan blog *Blogging Project Runway*. The trials and tribulations associated with their labor of transcription, and the existence of obstacles that ranged from computer crashes to car accidents, also found a place in the *Project Runway* fan blog.

In addition to offering a weekly podcast from the behind-the-scenes and yet in-front-of-the-camera fashion mentor Tim Gunn, the show's original network, Bravo, used a range of transmedia strategies to capitalize on the ubiquitous computing technologies and distributed online networks. In fact, ratings for the television broadcast appeared to have been driven partly by the show's popular website, which offered video clips of highlights and outtakes, show-themed AIM icons, and ringtones and wallpaper for cell phones.[26] Partnering websites such as Gay.com and Planetout.com were used by the show's marketers to further build on the online community base through a nexus of well-developed Internet subcultures.[27] The show also fostered participatory cultural practices by reaching out to a heterosexual media mainstream through a promotional online game called "Fashion Face-Off," which attempted to borrow the conventions of an "online football pool" from established practices of sports fandom, such as wagering over outcomes or making numerical predictions.[28] In mapping the complex transmedia genealogy related to the *Project Runway*, it is important to keep in mind how the podcast functioned in relationship to a number of other authorized restagings of the show's pedagogical drama on the Internet. These restagings included an eBay auction; a DIY site for video mash-ups; an online, interactive fashion-creation site where users could save designs made from virtual fabric swatches and generic separates; and several official blogs. Many of these online experiences encouraged visitors to engage in further rating activities, which included rating the designs and mash-ups of other viewers, even though procedurally the total number of possible outcomes and combinatory paths of creativity were extremely constrained.

One of the official *Project Runway* blogs was ostensibly written by Gunn himself, so that discriminating fans could conceivably have compared the rhetoric of the podcast to the rhetoric of his written blog, which was characterized by frequent parenthetical expressions, respectful but direct epistolary addresses to individual contestants about the success or failure of their work that particular week (which instructors would recognize as a classic teacher's comment paragraph), and seeming digressions that paid homage to the common Internet genre of the FAQ. The tones of the two rhetorical productions for Internet distribution were quite different. For example, in the blog post for episode 9 of season 3, Gunn tactfully said, "The designers quickly learn that Parisian fabric stores are very different from MOOD; specifically, they comingle fabrics for fashion and home furnishings, so one has to do a lot more digging." In the audio channel, Gunn's commentary was much more blunt: "I picked up the end of the bolt of this horrible fabric—I mean, it was like some bad upholstery fabric from some maiden aunt who lives in a cave ... Guess who chose it! Yes, he did. Actually, it looked better in the workroom than it did at the fabric store, but still, it looked like a couch."

It could be argued that the voice of Tim Gunn itself became a kind of character in the show, one that was always already seen as alienable from his physical person. By season 2, class-clown contestant Santino, who mimicked fellow designers Diana and Andre, decided to devote his prodigious mimetic vocal abilities primarily to Gunn. Although Gunn himself imitated the dialogue of contestants Vincent and Angela in season 3, he did so in his own voice rather than through appropriating the voices of others. In contrast, Santino cultivated a pitch-perfect copy of Gunn's voice. Hillel Schwartz has written about this kind of parroting and the destabilizing effect that it has on personae associated with authority.[29] However, what irritated Gunn wasn't Santino's imitation of his catch phrases and verbal tics, it was his original invention of scenarios. Like the fan fiction that Henry Jenkins describes in *Convergence Culture*, it is the use of copyrighted characters in unauthorized original plots that is perceived as most subversive to authorial systems of cultural control.

This unauthorized copying can be read in relationship to other forms of illegitimate replication that constitute drama in the show, such as when plagiarism has to be policed by administrative agents. In season 2, contestant Marla is publicly humiliated by the judges on the runway after she presents a black party dress very similar to one done by a name designer, a photo of which was given to her in the dossier about client Nicky Hilton. From the podcast, we learn that print materials are often seen as key enablers of prohibited forms of reproduction. In the podcast, for example, we learn that mentor Gunn expresses his considerable anxiety about Marla taking the dossier with her to the fabric store, although he claims not to realize at the time the potential for the making of illicit replicas. The podcast for season 3 provides even more information about a pivotal incident of plagiarism involving contestant Keith, in which he is expelled from the show after being found with contraband publications that include an illustrated history of fashion and a basic pattern book. From the podcast, we learn that the show's producers were contemplating their own forms of questionable doubling after Keith bolts from the set in shame. A Keith stand-in is used in group shots with the intention of—as Gunn says—Photoshopping his face onto the body of his doppelganger.

Certainly anxieties about spoiler communities are also alluded to in Tim's Podcast, often in relation to these acts of unauthorized reproduction. Gunn doesn't want Marla to take the dossier because he fears that it will fall into the hands of a disruptive fan, who will then be able to publicize the content of one of the show's episodes in advance. Gunn also reports that contestant Keith has violated the rules of the show by going on the Internet to defend his reputation, a process of inappropriate face-saving that would continue in response to displays by fans in websites of Keith's initial *Project Runway* portfolio that juxtapose its contents with precise tableaux from the runway shows of others.

In his work on fan cultures, "spoiling," and media convergence, Henry Jenkins has studied a number of "transmedia narratives," particularly those associated with reality shows, such as *Survivor* and *American Idol*.[30] Jenkins claims that audience participation and grassroots activism are fostered when stories are allowed to migrate to different genres, forms, geographies, chronologies, and media platforms from the original network television broadcast format for which they were intended.[31] However, *Project Runway* regulates fan behavior by policing its contestants and constraining the forms of permissible content-production taking place on its authorized, interactive websites. Although the podcast uses a one-to-many broadcast format, it often suggests a kind of potential complicity with fans. In Walter Ong's terms, like the "secondary orality" of the "telephone, radio, television and various kinds of sound tape," the secondary orality of a podcast may have what Ong calls a "participatory mystique" or "communal sense."[32]

The proprietary grasp of the show's producers on the products created by the labor, and even the goods, of others is also an important aspect of the podcast commentary, and the alienation created by the commodity capitalism that is elsewhere celebrated by the show's video and Web content is probed in Gunn's auditory performance. Although in season 2, Gunn says that he wants to "shake" would-be designer Kirsten for not being willing to use the fabric from an heirloom scarf in the "Clothes Off Your Back" challenge, by season 3 he wants designers to be cognizant of the ways that use of others' property literally can be considered theft on the show, based on the legal contracts into which they have entered. When contestant Angela discusses how she would like to incorporate the shawl of her "Everyday Woman" client into her look, Gunn explains his hesitation in the podcast.

"Lorraine will sacrifice that shawl. It will become part of a look. And it will be owned by the show. It's not going to be owned by her anymore."

In addition to his color commentary on intellectual property in the show, there are also counterdiscourses about class, gender, sexuality, and body image. Gunn's marked counternarrative is often oriented toward an oppositional stance against the judges. For example, in season 2, episode 4, Gunn talks about the "hidden sequestered feminist in me" who expresses surprise that the judges do not respond negatively to the *Playboy* heterosexist ideology of contestant Daniel V's winning design. In the ice skating challenge he implicitly questions the taste of the judges, particularly guest judge and former Olympian Sasha Cohen, in ways that betray certain subtle class ideologies by deploying words like "project" and "imagine," while undermining the judges evaluations of the "vulgar." In season 3, episode 5, Gunn discusses his private speculations about the sexual identities of the designers: "I never get involved with or even question the sexual preferences of our designers. I mean, we just do what we do, and we accept who we are, and we are who we are. But I thought 'Gee, I had sort of assumed that you were a gay male, but now I don't know.' How could any self-respecting gay male not know who Cher is!" In season 3, Gunn despairs over the lack of sensitivity of both the judges and the designers to the plight of fashion-conscious plus-sized women who can never live up to the fantasy proportions of professional models.

Strangely, there is a lacuna around issues of race in Gunn's podcast. The obvious fact that African Americans are competing for recognition as designers in an industry that assumes a culture of whiteness is never acknowledged in his counterdiscourse. In the first episode of season 2, when he talks about diversity, he is speaking about the variety represented by the contestants' professional experiences. Even code words, such as "multicultural" or "urban," that appear in the video broadcast of the show aren't alluded to in Tim's Podcast, although at one point he complains about how contestant Zuleima keeps to herself off camera. At the one point when race officially appears as an issue—in season 2's Barbie episode, in which designer Andre refuses to put a blond wig on his African-American model—the incommensurabilty of the person of color to the denotated "fashion icon" has to do with an object of gaze rather than a creator of a look, for which Gunn expresses no sympathy.

Unlike the potentially unreliable judges, when the producers appear as characters in his podcast, Gunn doesn't seem to question their authority. In season 3, episode 2, Gunn praises the managers of the panopticon for their ability to have omniscient knowledge that he lacks. While he often points out the judge's shortcomings in not identifying the signs of sloppy construction that take place behind the scenes, he also acknowledges the limits of his own gaze to fully comprehend the social dynamic between the designers and the superiority of the policing vision of the producers.

> It was a bitch slap, and a necessary one. I mean, our producers have the greatest integrity, they have the greatest perspective on things ... and I was getting very myopic. I was extremely grateful for them, and it was a good wakeup call for me. It was like pow! I did my own about face. It wasn't suddenly I was yelling at Angela, but it was a matter of ... I had a perspective now. It was like, okay there is more happening here. The producers through their own little minicam system know everything that is happening everywhere. I only experience what I experience. The producers, God bless them, don't fill my head with unnecessary stuff.

In some reality television shows, contestants do participate in the assumed omniscience of the producers, by seeing what is ostensibly hidden-camera footage and, in at least one case, by viewing polygraph results. The central trope of the earpiece and listening dematerialized in the podcast is even sometimes a covert element of reality TV interaction, as contestants are piped supposedly spontaneous dialogue through secret transmitters.

Gunn's podcasts both violate and reaffirm the validity of certain cultural taboos by associating his discourse with the academy and the Enlightenment values that it represents. It constitutes listening as a social act, but this sociality is one that comes with potential exclusions; this is particularly true regarding the divide between what he calls "design" and what he calls "dress making," which has now become a major issue as the DIY aesthetic is defended by those challenging the authority of designated experts in commercial design. Rather than project the intimacy of the office hour in his podcasts, Gunn treats us to the intimacy of campus gossip. He also provides the critical commentary that makes sense of the pedagogical drama of the show by emphasizing the faculty of judgment. He simultaneously delivers avuncular lessons about civic virtues, such as honesty and empathy, to be cultivated in the moral development of his student body in a process that seems separate from the activity of applying evaluation criteria to the work to be graded.

This complex range of rhetorical activities in the relatively banal venue of supplement to a reality TV show serves as a reminder that didactic and narrative structures can be quite compelling to audiences. The decision to make coursecasts in talking-head formats that are so antithetical to story and even to dialogue is one that should be reexamined before the forces of path dependence make it even more difficult to imagine alternatives. There is a lot of potentially interesting cultural work to be done, but distance learning often is more likely to resemble a traffic school tutorial rather than a compelling professorial performance. Obviously, reality television perpetuates the consumer capitalism, commodity fetishism, and niche stereotyping that feed the markets on which the entire economic model of broadcast media is based. Yet there is something to be learned by faculty and administrators from its highly engaging depictions of pedagogical drama.

Performance versus Pedagogy

Producers also play an active role in another media phenomenon, frequently starring academics, that is designed to appeal to life-long learners: the TED Talk videos that circulate through viral channels. The annual TED conference (TED stands for Technology, Entertainment, and Design) has expanded its popular offerings to include professors holding forth on subjects that include mathematics, astrophysics, evolutionary biology, neuroscience, and human sexuality. The most popular videos often feature research faculty from elite institutions, such as Harvard or Cambridge.

Although TED Talks seem to represent a radically new form of what Henry Jenkins has called "spreadable media" that mixes top-down and bottom-up distribution and fosters participatory culture at each junction of dissemination,[33] the most popular online videos are often those with conventional messages that follow traditional narrative structures and require an attitude of reverence from the audience. They may even borrow devices from much earlier in the history of American rhetoric. For example, brain scientist Jill Bolte Taylor's popular (having attracted over ten million views) and dramatic "Stroke of Insight" TED talk, which details her experience of suffering a catastrophic stroke and then recovering from it, borrows from much older popular rhetorics of captivity and conversion. Unlike captivity narratives about being held prisoner by supposedly uncivilized tribes or being enslaved by cruel masters, which were both popular themes in nonfiction bestsellers of the eighteenth and nineteenth centuries, Bolte Taylor, because of the failing left hemisphere of her brain, is a victim of captivity in her own body.[34] The rhetoric of disempowerment is critical to the rhetoric of empowerment that follows, when Bolte Taylor asserts that right-brain thinking allows insight into "the life-force power of the universe" and access to the "power to choose how and who we want to be in the world." As she copes with acknowledging her fallen state, she describes herself as an "infant in a woman's body" who "could not walk, talk, read, write, or recall" when deprived of her left-brain abilities. Later, she describes a classic conversion narrative in which she is transformed by the traumatic incident.

Much like testimony by those who reject a misguided past and achieve enlightenment through a profound spiritual experience, Bolte Taylor bears witness to the "peace" of embracing the "we inside of me."

Personal narratives with strong individual appeals to pathos often make for compelling TED Talks. Despite the framework of expertise in supposedly objective scientific research and the ethos of academic authority, subjective experience matters in that speeches given from the heart are assumed to be more authentic. Much like other Internet phenomena, the metrics of search-engine-driven keyword popularity also matter. There is even a TED Talk that pokes fun at statistical tendencies in TED Talk titles in which the speaker explains how a would-be speaker could game the system.

> Now, with the topic—there's a whole range of topics you can choose, but you should choose wisely, because your topic strongly correlates with how users will react to your Talk. Now, to make this more concrete, let's look at a list of top ten words that statistically stick out in the most favorite TED Talks and in the least favorite TED Talks. So if you came here to talk about how French coffee will spread happiness in our brains, that's a go. (Applause) Whereas, if you wanted to talk about your project involving oxygen, girls, aircraft—actually, I would like to hear that talk, but statistics say it's not so good. Oh, well. If you generalize this, the most favorite TED Talks are those that feature topics we can connect with, both easily and deeply, such as happiness, our own body, food, emotions. And the more technical topics, such as architecture, materials and, strangely enough, men, those are not good topics to talk about.[35]

If TED Talks are designed to further the continuing education of the general population, these patterns of consumption seem to signal a superficial approach to learning. What emerges from this word cloud of TED talk titles is information about the fundamentally narcissistic and hedonistic tendencies of the online audience and a widespread denial of the importance of infrastructure, material constraint, and gender politics that conflict with the more hopeful TED ideology of unfettered personal freedom. This propensity for telling the audience exactly what they might want to hear, and reinforcing an optimistic vision of the world, is so strong that it is worth noting that the number-one result that the TED search engine provides for the term "global warming" leads directly to the work of Bjorn Lomborg, who has become famous for minimizing climate change as a geopolitical priority.

To create a polished and compelling performance every time, a battalion of TED producers—or "curators" in the conference's parlance—is tasked with ensuring that recorded conference speakers don't hit any rhetorical wrong notes. The rigorous process of vetting candidates and rehearsing presentations to fit the formulae of previous successful talks is described in a

New Yorker article, in which writer Nathan Heller details how presenters are coached to select props and supplant verifiable statistics with autobiographical stories during "rituals of preparation" that focus on "emotional shading."[36] An article in *New York Magazine* is even more critical of the "schmoozefest" culture that dominates at TED and "the rise of a cohort of speakers and attendees who migrate along the same elite social-intellectual trade routes" and who reduce academic inquiry to the lowest common denominator of populism.[37]

Yet rhetorical flair is certainly important for effective instruction, and pedagogical drama definitely makes lectures more comprehensible and more memorable to students. As a former writing director of the Humanities Core Course at the University of California, Irvine, I watched hundreds of lectures designed for freshmen, and the best ones often used dubious tricks such as visual aids. I have seen faculty wearing costumes that ranged from an ancient Greek toga to a Jesuit habit and deploying props that included a can of malt liquor, a reproduction of an Aztec codex, and a cardboard cutout of a Starfleet commander. The drama of a live performance is important in a rhetorically effective university lecture, and more and more often students are expected to serve as performers as well.

When databases of pedagogical resources like those archived at the Khan Academy include TED lectures as representative of university-level arts and humanities instruction, however, it is worth asking if such presentations actually are good models for remote learning situations, and if their conventions should be emulated in the specific genre of the coursecast. TED Talks may present entertaining spectacles of the display of embodied disciplinary knowledge, but they also show a passive audience of spectators whose participation is highly constrained. TED also polices comments on its official website, thus inhibiting practices of participatory culture that could lead to deeper engagement and the potential for critique involved in real student empowerment.

The Hands of the Master

So how should a compelling coursecast be structured? We have considered formats that star students, such as *Project Runway*, and those that star instructors, such as TED, but we haven't said much about formats presenting disembodied instruction that relies on screen capture. What about the role of digital manipulation and sleight of hand in those presentations? Many universities have licensed commercial software products, such as Camtasia, or encouraged faculty to use free downloads, such as Camstudio, in order to use recordings made from a professor's computer screen to create coursecasts that show step-by-step how to find online resources, complete assigned work, solve scientific or mathematical problems, or learn to use complex software packages designed for artists, engineers, programmers, or other specialized disciplines. Ramsay discusses how screen recording by the typical faculty member often produces little more than a document of "fumbling" and rank amateurism, and I have observed that vernacular instructional video archived at sites like the Khan Academy may be even less polished, but it is worth noting that professionally produced coursecasts that are heavy on screen capture have become part of a growth industry in online training represented by companies, such as Lynda.com, that cater to the computer-graphics industry.

Like the TED Talk, these computer-graphics training videos borrow many of the rhetorical conventions of the tech demo aimed at Silicon Valley cognoscenti. For example, TED Talks have focused on how to use new products such as Photosynth or BumpTop. Their aim is to delight as well as to instruct with dazzling sleight of hand at key moments, although the tricks are always explained. Although Lynda.com videos may be addressed to beginners, the assumed access to expensive software makes these students automatic insiders with access to cultural capital. So it could be argued that the same type of elite software culture inhabited by the creative classes may be reproduced in both formats. Of course, the personal narratives that are so important to the TED Talks are absent in this much more technocratic delivery form of the tutorial, and the modularity of step-by-step organization is different from the continuity of the emotional arc at the core of the delivery of a TED speech.[38]

As I discovered while improving my last-century digital design skills at video editing and compositing classes at a local community college, public educational institutions often assign videos from the private Lynda.com library to save on distance-learning costs that might otherwise be prohibitive when measured in faculty time. In a prerequisite online Photoshop course that I was required to take, the contrast was notable between the videos that my professor produced and the slick Lynda.com videos, which featured dexterous and glib Photoshop stars such as Deke McClelland, Chris Orwig, and Jan Kabili. Some of my community college professor's videos appeared to be literally shot in his basement. Many included unedited moments of screen recording in which he accidentally performed the wrong operation or expressed verbal irritation with the results of his efforts.

The Lynda.com videos showcased what appeared to be very skilled live performers, although the step-by-step format of the course videos would have made the editing of particular shots relatively difficult to detect. Projects varied from doing quotidian exercises such as "Removing people from a photo" or "Rotating a pattern layer" to manufacturing fanciful grotesques like "Mapping a dog face onto a duck" or "Adding wings to a horse." Some Lynda.com celebrity instructors even make cameo appearances. The bearded, announcer-voiced McClelland appears in some of his videos wearing the headphone-microphone set that serves as part of his essential instructional kit. After a relatively brief introduction with his signature "Hi Gang" greeting, however, McClelland generally disappears from the learner's computer window. (One notable exception is McClelland's high-speed viral video "101 Photoshop Tips in Five Minutes," in which the Lynda.com guru vamps, raps, croons, and swears in an instruction-packed hyperlesson that no student is expected to keep up with.[39])

In thinking about how such screen recording coursecasts fit into the longer history of instruction, it may be useful to think about the tradition of *imitatio* and how apprentices may be trained to be experts by copying existing models.[40] Just as students of classical rhetoric were once instructed to emulate the specific features of famous speeches, sometimes by reproducing parts of the text piece by piece, Lynda.com students are expected to learn by duplicating, as they painstakingly reproduce a facsimile of the evolving work displayed by their instructors. Although these videos may lack the humanizing presence of either instructors or students in the frame, and the only faces on screen may be the static visages of stock photos, such modular course materials still may captivate attention. They also foreground the processes of digital labor and immerse students in the workflows of computational media as disembodied hands navigate menus, move sliders, manipulate layers, and trace pixels on the screen. Absent the embodied manifestation of the "sage on the stage" and given only the voice of "the guide on the side,"[41] the student is presented with a pedagogy that has both less and more authority. The teachers of Lynda.com lack the authority of the body, and they cannot direct the bulk and gestures of the human form in ways that command attention and respect, but instructional embodiment also creates vulnerability to challenge and critique. When Scott Klemmer's students sought him out in person to discuss what they had learned in his Human–Computer Interaction class, as the following chapter describes, their expectations of face-to-face mutual recognition might be very different from similar students enrolled in a Lynda.com course who only know faceless instructors.

The embodied college lecture is a genre that has evolved during the course of a long rhetorical history in the United States in a complex tradition that has been shaped by many influences. Different genealogies of the college lecture can be traced back to the ecclesiastical sermon, the scientific demonstration, and the general education public lecture typified by the nineteenth-century Chautauqua movement.[42] Because of the circumstances of its production and reception, the genre of the college coursecast borrows from other new media genres, while it also retains some elements of the traditional college lecture. Whether it is the "sage on the stage" of the TED Talk or the "guide on the side" of the arbiters of taste in reality television shows like *Project Runway*, college faculty have many more models to emulate than their nineteenth-century forebears. They also must consider how the interface of the Internet delivery system itself functions and how their online audiences are constituted when designing modular user experiences in the distance-learning situation.

5 The Rhetoric of the Open Courseware Movement

On a sunny summer day in a park in the heart of the Silicon Valley, a crowd assembled for a picnic. Burgers sizzled, sodas sloshed in bins of ice, and people circulated in and out of clusters of enthusiastic participants. But this was not a conventional company picnic, despite the presence of dozens of corporation employees milling about wearing blue "Coursera" T-shirts. This was a "meet-up" for a few hundred of the hundreds of thousands of students enrolled in Coursera's massive open online courses, or MOOCs, a grand experiment in free large-scale distance learning, in which there might be as many students simultaneously enrolled in one online course taught by a single professor as those studying in an entire state's public university system. Faculty marveled at suddenly having "about 200 years' worth of students" at one time thanks to Coursera.[1]

In the planning that led up to the picnic, organizers suddenly had to deal with managing many unfamiliar constraints for the Internet start-up: fixed location, parking and transportation, and the allocation of material resources that the virtual classroom had supposedly avoided. When Coursera's park reservation reached capacity, late RSVPs could not be accommodated, and rejected students behaved petulantly—either like irritated customers or like disappointed fans. Among the students who were able to attend, some responded to the picnic invitation with a strong desire to reciprocate. They brought extra food to make up for anticipated shortfalls. A table of potluck desserts included many home-baked treats.

Nametags indicated the titles of the college courses that students had signed up for. The class subjects included "Machine Learning," "Automata," and "Probabilistic Graphical Models." Students had traveled—sometimes across the country—to meet the professors that they only knew from their coursecasts, and by coming to the park students also finally had a chance to meet their fellow students. Many of them had carpooled to the park with strangers that they only knew via last-minute coordination on the Internet, but many others came with people they now considered online friends, after supporting each other through weeks of collaboration in their courses. Some had created T-shirts especially for the occasion. One pair of students wore T-shirts that read: "I am an expert in machine learning. Andrew Ng has declared it so."

This Northern California gathering represented only a small fraction of those who had decided to take advantage of the opportunity to study specialized subjects in free courses taught by Stanford faculty. As the *New York Times* reported, about two-thirds of Coursera's students encountered the course material from abroad. From Pakistan to the Philippines, thousands of students in the course I had taken had formed study groups with their fellow nationals to tackle the course material—which included video lectures, online quizzes, and peer-graded assignments and projects. These overseas students had clamored for access to translations and subtitled versions of the instructional videos. Many of them had requested that the picnic be live-streamed.

I had driven up from Los Angeles with a twenty-five-year-old female computer programmer who was originally from Belarus. She had completed "Algorithms: Design and Analysis" with Stanford Professor Tim Roughgarden. Despite the eclecticism of the group in the park, in several ways she was typical of people who attended the picnic, many of whom were already employed in high-technology companies or who were planning to enroll in graduate school in computer science-related fields. She had hoped to gain admission to the new Carnegie Mellon University campus in the Silicon Valley to specialize in software engineering, and she was immersing herself in U.S. higher education while she wrote programs to handle warehouse traffic for a specialty company catering to audiophiles. She was also an active participant in the "Russian Topic" forum for her class. At the picnic, she congregated almost exclusively with the other Russian speakers, who formed a large clique at the event. Many of them had computer science degrees from former Soviet Union countries, and they had been attracted to the easily recognizable Stanford brand and its association with high tech corporations.

I had taken the five-week course on human–computer interaction (HCI), which was taught by Professor Scott Klemmer, a well-known expert on HCI and computer interface design. The coursecasts of his lectures were full of interesting examples and stories about good and bad design solutions. We learned about the alternative histories of everyday objects, such as the measuring cup or the digital camera, and shared the professor's pet peeves about stupid design decisions in refrigerator coldness dials or conference registration pages. I was impressed by how well he adapted his professorial rhetoric

Figure 5.1
Stanford Professor Andrew Ng at the Coursera meet-up. Courtesy of Kimberly Spillman.

to this new genre, which was so different from lecturing to a live audience. I was also struck by the alacrity with which he recalibrated elements of the course in response to complaints on the forums that could have been exacerbated by his radical remoteness, the enormous scale of enrollment, and the diversity of the student body counted on to peer-grade projects.

Klemmer, encircled by his enthusiastic students, spent hours at the picnic dispensing advice and providing supplemental mini-lectures on subjects that ranged from introducing checklists to reduce medical error to simplifying complex click-through end-user license agreements. He also opined about pizza delivery systems, nuclear plant monitoring, radial computer navigation, mobile apps for finding campus parties, emoticons, avatars, and the unappreciated virtues of the QWERTY keyboard and Microsoft Office. There were many students with Klemmer who wanted help getting into graduate school, much like my carpool partner from Belarus, who discussed the application process with me during much of the long drive to the picnic. To a student who wanted guidance on writing the statement of purpose for her application for admission, Klemmer advised her to avoid anything that sounded hackneyed, such as "I've been fascinated with computers since childhood."

Although these kinds of massive open online courses have often been billed as a radical alternative to traditional higher education, what was striking about the interactions that I observed at the picnic was how many students had enrolled with Coursera in hopes of finally entering the orbit of American traditional higher education and even approaching the most rarified spheres of prestigious top-ranked research institutions. Many students wore clothing that showed the brands associated with a university affiliation. In the knots of Coursera distance learning students, it was easy to see Harvard hoodies and Berkeley tank tops, Carnegie Mellon T-shirts, and USC hats.

Klemmer not only fielded questions about writing personal statements for graduate school, he also was asked his opinion about soliciting good faculty letters of recommendation and choosing the most prestigious programs to guarantee professional success. In these discussions, Klemmer confessed to not really knowing three-quarters of the students in his live Stanford lectures for whom he wrote recommendations. He said he wrote hundreds of letters annually based on details provided by graduate teaching assistants and other records of student work. However, he was careful to personalize his address as he spoke with his formerly remote students, and frequently used their first names at the picnic, much as experienced professors try to memorize student names as quickly as possible by saying them out loud frequently.

Certainly, some students at the picnic clearly rejected the constraints of higher education, like the programmer wearing the T-shirt that read "I Piss Excellence" who confidently showed Klemmer the app that he designed on his iPad. This twenty-something self-proclaimed autodidact bragged of avoiding the limitations entailed by study for a college degree with the same pride with which he described building his own house. There were other stories of DIY learning floating among the conversations at the picnic—such as the girl who learned to braid her hair on YouTube, the man who learned photography online, and the mother who learned to fix an audio speaker system on the Internet; but students who expressed actual distain for the traditional classroom were obviously in the minority. Many were approaching the professors present timidly and primarily to talk about anxieties about their own perceived lack of achievements in conventional learning institutions, at least in institutions that would be recognized by the U.S. gatekeepers that controlled access to graduate education and elite research communities.

It was definitely Coursera's party, but many of the students present had also sampled the wares of the company's open education competitors. A female high school student who had come to the picnic with her father had started her online studies with the Khan Academy and had translated some of the videos there into French as a service to her fellow learners. The Khan Academy was started by a Silicon Valley investment fund analyst who had started tutoring younger cousins in math via YouTube in 2006. At the time of the picnic six years later, the Khan Academy could boast of having 3.5 million monthly users around the world and the attention of major funders such as Google and the Bill & Melinda Gates Foundation. Plans were underway to expand the infrastructure of the Khan video library to better track the individual progress of students and to introduce badges and other reputation systems. These "badges worth bragging about" on a site "full of game mechanics" supposedly would challenge students pursuing "bragging rights" as well as academic knowledge.[2] Intrepid researchers in cognitive science at the University of California, San Diego have even customized Khan Academy materials to create a "KA Lite" download that allows students in developing countries who don't have an Internet connection to access instruction offline.[3]

A college instructor munching on his burger praised Udacity as a better alternative to Coursera, right to the face of one of his blue T-shirt wearing hosts. He was referring to the company that emerged out of one of one of the most famous pioneering experiments in radically open education, the Stanford University Artificial Intelligence course that was offered online to over 160 thousand students from 190 countries in fall 2011.[4] On the original AI course page, which is still topped by a banner illustration of a bald-headed humanoid with cyborg vision, the class is described as a "bold experiment" that would explain the technical principles behind the latest technologies, including "Google Goggles, self-driving cars, even software that suggests music you might like" with similar "materials, assignments, and exams" as those in Stanford's introductory AI course.[5] The class was cotaught by Sebastian Thrun and Peter Norvig.[6]

In the opening video for the AI course, Thrun and Norvig promised that students who completed the more rigorous, homework-based version of the course would be provided with a signed "letter of accomplishment" and given information about their ranking in the class.[7] The professorial duo also provided "office hours" videos in which they answered student questions that were ranked by the voting system of Google Commentator. In their first office hours video they reported receiving over 300 initial questions and 12,000 votes. The top two, very similar questions appeared to reflect students' urge for more advanced courses in the traditional building-block model of a conventionally linear university program of study: "Is there a chance that you would continue with an AI II class?" from a Russian student and "Does Stanford or Google have plans to make more courses available to go deeper in the understanding of AI and getting our hands into applications?" from a student in Spain.[8] Thrun and Norvig also responded to complaints about privileging math over programming and provided encouraging tips about open source and DIY approaches to their subject matter.

Udacity, the spin-off company that now hosts the Thrun–Norvig AI course, claims to eschew "long, boring lectures" in favor of "solving challenging problems and pursuing udacious projects with world-renowned university instructors."[9] The company emphasizes four elements in its branding: (1) a library of free courses, (2) a community of learners and teachers, (3) the option of certifying course skills at a testing center, and (4) the option of making a consenting student's résumé available to one of twenty corporate partners recruiting new employees. The third element of the Udacity plan distinguishes the company from its close Silicon Valley/Stanford competitor Coursera, which was embroiled in a plagiarism scandal in August of 2012 involving its humanities courses. At the Coursera picnic, students were grousing about the possibility of class cheaters, although many still expressed confidence that a certificate of accomplishment from Coursera would be an asset in the workforce. Both job seekers and hiring managers spoke of the value of the certificates, although almost no students that I spoke with at the picnic had earned them by finishing all the projects.

Toward the end of the Coursera picnic, Machine Learning professor Andrew Ng addressed the crowd with an appeal to higher ideals about what it means to find a "real world solution that doesn't have anything to do with computers." He also evoked the memory of "the day you worked the hardest." As the separate cluster of "Russian Topic" shrank to only a few individuals, and the park began to empty, those who had studied course materials from abroad and families devoted to home schooling and "unschooling" lingered with the last stragglers. Cofounders Ng and Daphne Koller chatted informally with these diehard enthusiasts. Topics ranged from their attitudes about honor codes and academic integrity to their willingness to accept the résumés of former students and their schemes for generating revenue in the future. The rhetoric was one that emphasized expansion, both in course offerings and in participating institutions. The *Chronicle of Higher Education* had recently reported that college administrations, which normally move at a "glacial pace" toward new technologies, were "rushing into deals" with Coursera, which the newspaper characterized as an "upstart company" that was currently relying on blue sky "brainstorming" rather than a conventional business plan.[10]

At the picnic, Ng and Koller explained the history of Coursera by pointing to its origins in Stanford's Engineering Everywhere program, an initiative that received financial support both from the university and from the high-tech investment firm Sequoia Capital. Unlike Coursera, Engineering Everywhere prominently displayed a Creative Commons license and publicized a commitment to "free and open use, reuse, adaptation and redistribution."[11] Intellectual property was obviously a thorny issue where Coursera was not yet clearly positioned. At the time, Creative Commons licenses appeared on the landing page of the Khan Academy, Open Yale Courses, and Engineering Everywhere. The home page of Udacity and edX—where materials from the MIT, Berkeley, and Harvard brands were housed—used the traditional copyright symbol, although the latter mixed language with a "some rights reserved" disclaimer that provided a nod to more open licensing.[12]

By June of 2013, Coursera had expanded its Stanford/Michigan base and was listing over seventy institutions of higher learning, including Ivy League campuses Princeton and Penn and technical innovation hubs Cal Tech and Georgia Tech. Yet during this rapid expansion phase there was still uncertainty about which intellectual property regime Coursera would be ultimately using or how materials would be licensed in future. Its contract with the University of Michigan indicated that rights would remain with the individual faculty member "and/or" with the university, and the university was also responsible for copyright clearance that could involve third-party claims.[13] The conditions of this contract might thereby raise concerns not only about continuing open access but also about academic freedom and fair use. Other contractual obligations further complicated the status of content providers. For example, the University of Pennsylvania admitted that the institution's regular privacy policies that protected students would not cover Coursera users in Penn's online courses.[14]

Under the pressure of public scrutiny, Koller had to become an experienced public speaker in the spotlight within a short period of time, as the MOOC phenomenon quickly became an even larger news story. She soon appeared in a prominent TED Talk, which rapidly received over a million views, with an emotional appeal that contrasted her own fortunate position as a "third generation PhD" with "unfortunate" people in other "parts of the world."[15] In the video Koller then narrates a series of heart-wrenching tales of scarcity in higher education: a mother in South Africa trampled in a stampede for college slots, a student in a small town in India who would otherwise never have access to a Stanford education, a single mother in pursuit of social mobility, and the father of a patient with an immune deficiency disease. The talk was later deconstructed by a critic who noted that "Koller's vocal delivery is oddly modulated and arrhythmic, suggesting a memorized speech and not a natural delivery" and that "likely her array of vignettes has been vetted by focus groups who, like the analytics she also refers to, had a hand in deciding which sub-narratives carried the most emotional power" to produce "a carefully calibrated infomercial, one that has been created specifically to push all the right buttons."[16]

Because over 80 percent of Coursera students already hold undergraduate degrees,[17] it may be understandable to accuse the company of being disingenuous in claiming that its primary mission is helping the disadvantaged excluded from higher education. But Coursera has not been alone in making such appeals with a missionary mentality. In an editorial for the *Guardian* illustrated by a stock photo of a woman of African descent at a computer keyboard, edX president Anant Agarwal promises to "make education borderless, gender-blind, race-blind, class-blind and bank account-blind,"[18] and the top result for "picture of the day" for the Udacity blog shows students at the open university in West Ghana.[19]

The visual rhetoric of need, anticipation, darkest Africa, and American techno-salvation for people of color is difficult to miss. In the One Laptop Per Child initiative of the previous decade, as Rayvon Fouché points out, rhetorical appeals with problematic assumptions were likely to "embody historically constituted American racial politics and transport these politics to the developing world."[20] In lauding Fouché's research, Lisa Nakamura and Peter Chow-White observe that technoscientific initiatives based on the deprivation model are characterized by the belief that software can "transcend region, language, space, and culture."[21] Another promoter of Fouché's work, Siva Vaidhyanathan, contends that the latent colonialism of universalizing missionary work is typical of companies like Google,[22] which has also entered the MOOC market with its Course Builder application.[23]

Georgia Tech professor Ian Bogost has argued that the rhetoric of MOOCs has much more to do with the ambitions of the Silicon Valley and the anxieties of the Ivy League than it does with the aspirations of the developing world.[24] From the perspective of universities, he argues that alliances with MOOCs are (1) "a type of marketing" to promote particular brands to naïve consumers, (2) "a financial policy for higher education" derived from its asymmetrical participation in "disaster capitalism" that leaves universities unable to find alternatives in times of crisis, and (3) "an academic labor policy" intent on deskilling the work of instructors who have lost the power of faculty governance in increasingly more entrepreneurial schemes. Bogost argues that from the perspective of high-tech venture capitalists, MOOCs represent (1) "speculative financial instruments," (2) "an expression of Silicon Valley values," and (3) "a kind of entertainment" for the post-broadcast era of computational media, much like the TED Talks at which MOOC founders excel.[25]

Fred Turner tells the story of how the "Californian ideology" that combines libertarian and hippie thinking has informed a variety of cyberutopian initiatives over the past several decades.[26] If the software philosophy of the Golden State, which mixes entrepreneurial individualism with communitarian support networks, has shaped the rise of neoliberal MOOCs, state institutions that compose the higher education public infrastructure of California have often manifested significant resistance. For example, philosophy professors at San Jose State published an open letter to Harvard professor Michael Sandel expressing their outrage at being coerced to assign his edX "Justice" lectures.[27] The state university professors complained that a course in social justice required more intimate scales of personal engagement to facilitate putting theory into practice, and they also objected that the "digital generation" needed books rather than videos to promote their intellectual development through meaningful challenges. They also expressed concern that this policy turn toward uniformity and standardization shaped by elite institutions would drastically limit the academic freedom that promotes diversity of opinion.

> The thought of the exact same social justice course being taught in various philosophy departments across the country is downright scary—something out of a dystopian novel. Departments across the country possess unique specialization and character, and should stay that way. Universities tend not to hire their own graduates for a reason. Diversity in schools of thought and plurality of points of view are at the heart of liberal education.[28]

Sandel published his own letter to the San Jose State faculty supporting their central contention that "online courses are no substitute for the personal engagement of teachers with students, especially in the humanities" and acknowledged the "legitimate concern" of his professional colleagues in California about "budgetary pressures."[29]

It is important to remember that the often contentious relationship between distance learning and California state government has a history. In a 2005 paper published by Berkeley's Center for Studies of Higher Education, I examined four online collaborative efforts from 2001 that were designed to improve the quality of composition instruction that undergraduates receive by using variations of what would later be called the courseware model: MERLOT, CPR, UCWRITE, and SPIDER. All of these projects received public funding, all consequently manifested aspects of the digital politics of "the virtual state," and all originated with faculty serving large and culturally diverse student populations in California. Ultimately, they all also failed to achieve their goal for widespread cross-campus adoption and are now mostly offline, undersubscribed, or being phased out. Although these projects were designed for face-to-face instructional communities rather than the distance learning envisioned by many open courseware advocates, they illustrate the difficulties faced by programs that depend on an ethos of openness. Many of the same pedagogical concepts about using distributed computational media that would reappear in the position papers of the open courseware movement—intelligent tutoring, peer-to-peer exchanges of information, the vetting of material through participatory-culture rating systems—had been rehearsed earlier in California. Many of the same individuals were even involved.

When California state legislators put forward SB 520 in February 2013 to allow the recognition of MOOC credits if required courses at conventional public campuses lacked adequate seats to support enrollment needs, faculty rebelled. The bill was eventually amended to emphasize voluntary incentive grants rather than mandatory outsourcing to MOOC platforms,[30] but the damage to faculty trust was already done. The University of California Academic Senate wrote a letter that registered, in strong terms, their feelings of betrayal about not being consulted when the legislation was drafted. These public university professors protested diverting more state funding and accelerating privatization. They suggested that rigor meriting accreditation, student access to classes, and required time to degree might be better at UC institutions than at comparable MOOCs despite limited state resources for public instruction.[31]

Reengineering the DNA of the University

The University of Virginia (UVA) joined the list of Coursera participants after a contentious battle over online course initiatives that shook up UVA's leadership in the summer of 2012. The drama peaked when the university's president Teresa Sullivan was publically deposed and then reinstated.[32] Behind the scenes were acolytes of Clayton M. Christensen—a Harvard Business School professor—and Henry J. Erying—an administrator at one of Brigham Young University's campuses. According to the *New York Times*, the Christensen and Erying book *The Innovative University* was required reading among the entrepreneurs who tried to engineer a "campus coup" to oust Sullivan. The university's rector Helen Dragas had praised Christensen's model of "disruptive innovation" and thought that Sullivan was hindering needed progress toward radical change. Faculty responded with outrage. Most prominent among them, distance learning critic Siva Vaidhyanathan defended Sullivan in *Slate* magazine, calling the plotters on the university's board "robber barons" and denouncing the influence of "the rhetoric of Silicon Valley and the finance culture that supports it."[33]

The Innovative University employs the verbal and visual rhetoric of genetic engineering to make its case and promises a plan for "changing the DNA of Higher Education from the Inside Out." The double helix appears on the cover of the book, as well as on the sidebars for specific case studies. According to the introduction, the authors chose the genetics metaphor to contrast the DNA of BYU-Idaho with the DNA of Harvard and to show "how other institutions could change their DNA." Much as this biochemical code dictates formation of the components in a cell, university DNA shapes "the structures of departments," "the relationships among faculty and administrators," the dictates of "course catalogs," "standards for admitting students and promoting professors," "strategies for raising funds," and the plans for "campus buildings and grounds."[34]

Like corporations, the institutional DNA of universities might interfere with short-term and long-term survival. For example, in the survival-of-the-fittest battle between General Motors and Toyota, the American car company's fixed directive to manufacture ever-larger vehicles left it vulnerable to its more nimble Japanese competitor. Fortunately, according to the authors, visionary architects could redesign an institution's genome, as mathematician Isaac Greenwood did in altering the genetic blueprint of the original Puritan college to transform Harvard into a modern research university. In the nineteenth century, chemistry professor Charles Eliot, the "father of American higher education" in the opinion of the book's authors, would splice electives into Harvard's DNA, and student choice would cause Harvard to eventually shed its Greek and Latin requirements through the natural processes of disuse. In the twentieth century, organic chemist James Bryant Conant introduced more meritocracy to the campus to enhance the institution's competitiveness and nurture a better-adapted student body.

Of course, anyone with a high-school-level knowledge of genetics would have problems with the DNA metaphor. Genetic engineering can have unanticipated consequences, evolution is not always adaptive, and recent discoveries about the function of what was once considered junk DNA show the hubris of making inferences about the relationship between genotypes and phenotypes based on rudimentary theories and incomplete evidence.

Furthermore, genetic engineering—like other medical sciences—is a complex and at times messy enterprise in which inconsistencies in practices, policies, and procedures need to be negotiated by participants. In their classic book on craftwork and tacit knowledge in the field of reproductive medicine, *The Right Tools for the Job*, Adele Clarke and Joan Fujimora assert that "tools," "jobs," and "rightness" are all situationally constructed.[35] According to Clarke and Fujimora, workers must constantly engage with ad hoc arrangements and doable problems, as well with disciplining tools. The labor of reproductive medicine requires enrolling allies, analyzing texts of inscription and representation, and participating in recognized practices that adapt to constraints, opportunities, and resources. Feminist science and technology scholars have long questioned the instrumentalism and faith in standardization that is at the heart of Christensen and Erying's argument, and they have done so based on observations done in actual labs.[36]

Moreover, the authors' absolute faith in top-down executive decision-making seems potentially disastrous to anyone familiar with recent economic meltdowns and hubristic speculation driven by corporate CEOs who saw themselves as godlike creators of new business models. At least antiuniversity abolitionists believe in the wisdom of crowds rather than the wisdom of individual free-market technocrats. Jeff Jarvis and Anya Kamenetz may praise Wikipedia uncritically without acknowledging the vulnerabilities represented by the site's revert wars, uneven coverage, and shaky plan for economic sustainability, but they don't believe that a single individual should be able to redesign the entire platform of the world's largest reference work alone. Given the prominence of Wikipedia as a case study in other books about transforming higher education, to the extent that it sometimes appears in book titles on the subject,[37] the absence of such models of collective knowledge-making in *The Innovative University* is particularly noteworthy.[38] The rise of new practices around participation, viral dissemination, and free media that are such central concerns for other reformers apparently have nothing to do with the "disruptive innovation" envisioned. Ever the technocrat focusing on centralization, Christensen credits the digital university to the existence of high-speed, large-bandwidth infrastructure and the political influence of two western governors who pushed for new accreditation rules.

The Rhetoric of Experiment

Certainly, the rhetoric around MOOCs was often hyperbolic and declarative during its launch period. The *New York Times* called 2012 "the year of the MOOC,"[39] and the *Chronicle of Higher Education* announced that it was "the year of the mega-class."[40] *NYT* columnist Thomas Friedman devoted multiple columns to the "revolution" supposedly shaking the foundations of traditional education.[41] Yet there has been surprisingly little empirical study of the student experience in MOOC education and a paucity of independent research more generally. The Bill & Melinda Gates Foundation launched the MOOC Research Hub to analyze the results of MOOC experiments more systematically so that "researchers, academics, administrators, learners, and policy makers" can explore a range of questions as to the effectiveness of this format of teaching and learning, because to date "the impact of MOOCs has been largely disseminated through press releases and university reports" rather than peer-reviewed research.[42]

In other words, what many of these pronouncements in the media about trends for the future of education miss is the fact that these projects are experiments—in which failure is a distinct possibility and understanding precedents is important. To give the founders of Coursera credit, at least they understood their identity as a high-tech start-up that, rather than reinventing the wheel, capitalized on existing technologies, user behaviors, and infrastructures. Koller and Ng acknowledged the fact that risk taking was involved and frequently predicted that they would look back on innumerable mistakes based on problems already encountered.

Within months of the official launch of the company, Coursera students were already receiving apologetic e-mails describing how course pages had gone down when Coursera's commercial hosting company, GoDaddy, was hacked, explaining that there were "technical glitches" that were caused by "pushing out a lot of code within a short time frame" and "unexpected code interactions that have led to bugs."[43] Problems with student retention are common in online education, but Coursera has also sometimes struggled with problems of faculty retention as well. One University of California, Irvine instructor abandoned teaching his Coursera course on "microeconomics for managers" after a dispute in which the professor announced that he would never cave on his standards and cautioned that "any statement of accomplishment will not be worth the digits they are printed on" if he should.[44]

Perhaps the most famous failure braved by Coursera to date was what one reporter called the "MOOC Mess"[45] that occurred when a course called "Fundamentals of Online Education: Planning and Application" self-destructed spectacularly. Instructor Fatima Wirth lost complete control of the class when she attempted to facilitate small group discussions. Although she suggested that these discussions could take place on a "platform of your choice, Google+, Facebook, Skype etc." to allow for flexibility to accommodate the needs of the group, she initially organized students with a single technology—Google spreadsheets—that turned out to be unable to handle such a huge volume of users working with a single document. Soon students were deleting and vandalizing entries. To make matters worse, tabs didn't work, videos didn't run, and instructions for assignments were missing.[46]

Some said such a catastrophic failure was inevitable. In her introductory Coursera video, Georgia Tech instructor Wirth identified herself primarily as an "instructional designer" rather than as a faculty member.[47] Without question, she lacked the authority of research faculty teaching other courses, and she seemed uncomfortable as she spoke stiffly in front of a purple curtain. After the course was abruptly canceled, mockery of the professor on the forums leaked into the blogosphere. Students criticized her "mind-numbing laundry lists of PowerPoint bullet points" and observed that Coursera needed "performers who can deliver riveting lectures."[48] What made the disaster particularly mortifying was the fact that many of the 41,000 enrolled students were "professors, teachers and experts on online education."[49]

Fortunately, having large numbers of instructors enrolled in common courses has facilitated considerable discussion about pedagogy. Large numbers of educators also signed up for "E-Learning and Digital Cultures" with Coursera. Because the instructors of this course focused on using low-bandwidth and low-interactivity commercial social media technologies, such as Flickr, in ways that they were confident were sustainable, they were able better to handle technical glitches, but students still publicly aired their discontent with basic course design. Some argued that the course emphasis on contrasting utopian and dystopian attitudes about technology was too simplistic for a college-level course aimed at advanced students who might be aiming to create such course experiences themselves.[50] Others observed that the unstructured format, designed to validate a variety of viewpoints, and the abstract and general learning objectives created an experience more like a "happening" than a course and was, thus, difficult to assess.[51] Many expressed skepticism that a class that set the bar so low and presented such thin tools for assessment could grant college credit that would count as units toward a degree.[52]

The audience for the e-learning course quickly perceived that there were trade-offs when access and participation were prioritized over other academic values. However, the relationship between student motivation and curricular challenge is a complicated one. In examining the psychology of optimal experience, Mihaly Csikszentmihalyi famously maintained that "flow" was facilitated most by an inverted-U design—one in which systems had few barriers to basic access but many challenges in keeping interactions interesting and engaging, motivating the building of expertise, and, eventually, spurring the acquisition of mastery.[53] Although some dispute Csikszentmihalyi's hypothesis based on their own experimental observations in which people seem to engage most actively and for the longest periods with learning technologies oriented around easy successes,[54] MOOC courses, even those with relatively unchallenging subject matter, tend to rack up dropouts quickly. Although Koller and Ng have argued that the very low retention rates of MOOC curricula—hovering near 5 percent—should not be considered a problem,[55] most educators consider retention an important indicator of the effectiveness of teaching and learning.[56] In traditional classrooms the expectation is that students who begin a course of study will finish it; enrollments must be monitored to ensure that finite resources are measured out effectively and fairly. Koller and Ng dismiss this narrow focus on conventional matriculation as ignoring "life goals" and "learner intent."

Koller and Ng's vocabulary deliberately moves the rhetorical focus from *teaching* to *learning* and often appropriates the lexicon of the "connected learning" movement, despite the company's proprietary interest in the hierarchical structure of MOOCs. Connected learning articulates a number of learner-centered principles that reorient the power relations of the conventional classroom. The connectedlearning.tv website, which hosts relevant videos and forums, defines connected learning as "when you're pursuing knowledge and expertise around something you care deeply about, and you're supported by friends and institutions who share and recognize this common passion or purpose."[57] An FAQ explains that "sports, creative activities, intellectual pursuits, civic action, or games" could be included in "connected learning-type" experiences, and the emphasis is on lifelong learning rather than the traditional learner.[58] However, connected learning advocates explicitly distinguish informal learning communities from particular technologies, tools, or techniques. They also affirm commitments to "social equity and progressive learning" and "long-standing values around learning that have to do with relevance, social connection, and linking connections to opportunity and the real world" that seem at odds with the corporate culture of many MOOC start-ups.[59] Some prominent advocates for connected learning authored "A Bill of Rights and Principles for Learning in the Digital Age" and posted their declaration on GitHub, as though their manifesto insisting upon guarantees to privacy, transparency, data ownership, and reciprocity in online learning to counter potential commercial tyranny were lines of computer code contributed to the popular software programming community.[60]

The potential for comprehensive corporate surveillance that generated such anxiety among connected learning champions was celebrated by Coursera's promoters as a triumph for evidence-based methods. Because of the large number of students feeding material into information streams that are scrupulously logged by computers, it becomes possible to look at discrete student learning behaviors with a "big data" model that hypothetically could generate real research questions about how to teach more effectively in the digital age. For example, Stanford education professor Dan McFarland tested how students felt about the presence of the speaker potentially distracting attention from information of slides by creating two iterations of his course: one in which he appeared as an embodied presence and one in which his disembodied voice only narrated the slides.[61] Students, it turns out, preferred to see their professor as a sage on the stage rather than a guide to the side, and this experiment in A/B testing was quickly called off.

The diversity of the online student body could also facilitate new ways to model faculty interactions. Princeton sociology professor Mitchell Duneier hand-picked students from around the world to participate in intimate seminar experiences using video-chat technologies, and he also invested time in individually grading selections of work to refine peer-review procedures.[62] Many faculty have spoken of the virtue of flipping classrooms—by outsourcing impersonal lectures—so that more class time could be devoted to mentoring project-based work. For example, Duke professor Mohamed Noor assigned the lectures that he made for Coursera's "Introduction to Genetics and Evolution" as homework and changed his in-class lesson planning to emphasize active practice and the application of theory once learned passively.[63] Ironically, Duneier defected from Coursera after hearing that his course materials might be used by other instructors in flipped classrooms. As a sociologist committed to inclusion and government support for education, he feared his participation could unintentionally become "an excuse for state legislatures to cut funding to state universities."[64]

A Jury of One's Peers

Experiments in peer grading and peer evaluation have become an important part of the latest wave of distance learning. Obviously no professor—even with an army of TAs—can address the individual needs of tens of thousands of remote students, so peer feedback is the only way to address the learner's desire for engagement with a real audience. Despite a desire for supposedly labor-free machine grading systems, course management platforms like Coursera depend on human input. Attempts to introduce "robo-grading" of essay assignments that rely on written expression risk public mockery,[65] particularly when critics like MIT's Les Perelman point out the fallibility of supposedly intelligent systems.[66]

Peer evaluation is really nothing new: it predates the self-help movement in the United States by centuries.[67] As Lynée Lewis Gaillet points out, George Jardine—a professor of logic and philosophy at the University of Glasgow from 1774 to 1826—developed techniques for peer review to assist underrepresented and underprepared students and to provide more scaffolding for their completion of assignments.[68] Yet many are justifiably suspicious of neoliberal regimes propagated by so-called "mechanical turk" activities that harness cheap online crowsourcing technologies to facilitate finer distinctions and more accurate inferences than computers can currently make.[69] (The term "mechanical turk" comes from an eighteenth-century chess-playing automaton operated by a human being.)

Unfortunately, the promises of telepresence and ubiquity can be difficult to keep. Although advocates for participatory culture, such as Cathy Davidson, push for more peer grading to transform the hierarchies and one-to-many communication modes of the academy,[70] the issue of how specific software packages for peer review function as media experiences that involve human–computer interaction is often overlooked. Coursera is currently experimenting with a system called "Calibrated Peer Review" (CPR) as a way of providing assessment mechanisms of more complexity than the multiple choice tests on which Coursera often otherwise depends. For example, a Stanford professor's open-book final examination in his "Introduction to Mathematical Thinking" course is graded with a calibrated peer review system.[71] CPR boasts of National Science Foundation support and points to tens of thousands of student accounts as evidence of its credibility; it is now vying for greater adoption around the world with the aim of relieving faculty members of the labor of grading individual students and encouraging more writing assignments in large enrollment courses.

CPR students move through three basic hierarchies of competence when using the online interface: instructional evaluation, peer evaluation, and self-evaluation. They begin by learning to apply evaluation criteria to writing samples that have already been scored by the instructor; they then move on to assigning scores to "live" papers written by their fellow students. To keep their calibrations from becoming inflated, students are rewarded for accurately estimating how their own compositions are likely to be graded by peers. As faculty who use the system readily admit, students can be punitive about errors in language use that instructors might read more generously as they evaluate the mastery of subject matter over the form of expression.[72] Furthermore, exceptional students who attempt more ambitious responses to an assignment may be marked down by literal-minded students less able to recognize creativity or innovation.

I first wrote about CPR in 2005 when I was evaluating a number of different online approaches to improving writing instruction in public higher education. At the time I was serving on an advisory committee concerned with "Teaching and Learning with Technology" for the ten University of California campuses; I assessed how various writing-instruction software programs and websites compared to best practices in teaching composition. Some of these initiatives seemed promising, but I doubted that student writing for CPR would improve as rapidly as it does when writers have a clear sense of audience and purpose.

> One could say that CPR is all system and no lifeworld, communicative action with no second person discourse, an automated series of assignments that the instructor never actually reads. Although CPR ostensibly emphasizes a structure of "peer review," the review process is aimed at quantitative activities of calibration totally divorced from rhetorical context. ... Unlike peer-to-peer networks that encourage online communities by facilitating spaces for chat and messaging, the atomistic organization of CPR discourages students from occupying positions of agency either with each other or with the instructor.[73]

The creators of the CPR system—led by University of California, Los Angeles chemistry instructor Arlene Russell—claimed to combine "the pedagogy of 'writing-across-the-curriculum' with the process of academic peer review."[74] Yet research about online communities, best practices for creating effective user interfaces, and the empirical study of large samples of student writing were not reflected in the design philosophy of the site.

Unlike many of the online writing initiatives that I evaluated in 2005, in 2013 CPR was still in business, with an active website and expanding user base. The Regents of the University of California still controlled copyrights and trademarks on the CPR site. Although there was a page of publications with articles about the CPR system's virtues from journals such as *Advances in Physiology Education* and the *Journal of Undergraduate Neuroscience Education*,[75] many of the studies lacked the rigorous methodologies demanded in scientific fields. For example, although boosters trumpeted a "10% improvement" rate,[76] there were no research studies that actually compared the progress of CPR students giving and receiving calibrations to students in a control group who might have only seen the more articulated prompts, rubrics, and samples developed by the faculty member for the CPR training purposes.

Focusing on clarity and specificity in prompts and rubrics, as well as providing numerous samples of model responses, are pedagogical approaches long-recognized as best practices in composition. Sadly, the CPR prompts for feedback from students did not apply current research indicating that common instructional comments—such as those about "flow" and "awkward" phrasing—may be too ambiguous to change writers' behavior; nor were findings applied from large-scale longitudinal studies of college writing, such as Nancy Sommers's famed Harvard study of student writers,[77] to optimize feedback.

Nonetheless, I would not want to demonize those who use the CPR system. Many of the CPR early adopters were faculty members committed to student learning. For example, the CPR system was used by Ed Hutchins, a professor of cognitive science and a noted researcher on embodied interaction who introduces students to research on distributed cognition in case studies, such as groups of pilots working together in cockpits. Hutchins' work on the complexities of "thinking through the body" and coordinated, live real-time behavior could also be applied to the writing situation of his own students, who were producing discourse with various people using differing technologies in constrained places and spaces. Hutchins sincerely wanted to see his students' writing improve, yet he didn't initially think to study how they might improve while working with low-tech technologies and with each other outside of the constraints of the CPR platform.

Issues about embodiment or affect are too often absent in planning for serving students remotely, even though these issues may be incredibly important for how participants experience copresence online, how they articulate the norms of computer-mediated communication, and how they interpret and understand their own educational histories. Seemingly banal status updates—about eating burritos or listening to incidental music—that are common on social media sites and microblogging platforms may seem relatively unimportant as social rituals, but these everyday messages convey important information about how digital subjects think through their bodies and represent their own corporeal presences in time and space.

Feelings of copresence and of willingness to disclose personal information to others can be intensified by particular strategies that make the live qualities of remote interactions more salient. For example, Stanford researchers have studied how using avatars may facilitate more robust exchanges in collaborative virtual environments; they found that avatar use may sometimes even have advantages over traditional teleconferencing in situations where it might be otherwise difficult to experiment with new social roles or transmit highly intentional behavioral cues.[78] The phenomenon of simulating copresence with teachers and peers seems to be critically important for online learning as well. Yet, just as feelings about more than face-to-face interactions with other human beings shape educational experiences, many different aspects of human–computer interaction can influence the learning situation. Nonetheless, researchers in the Affective Computer Lab at MIT observe that the role of emotion is frequently discounted, even though it is "fundamental to human experience, influencing cognition, perception, and everyday tasks such as learning, communication, and even rational decision-making."[79]

As faculty members and students experiment with radically open course formats and flipped classrooms, they must negotiate the challenging territory between the high expectations set by the optimistic sales pitches of course promoters and the depths of the flame wars instigated by disaffected networked actors. A collection of comments aggregated by @attackcomplex documents "Flipped Class Hate Tweets" from dozens of dissatisfied students around the country; in posting to Twitter many expressed their vitriol about flipped classrooms with four-letter words.[80] Students "shouted" by using capital letters with criticisms that included "If you're a TEACHER then TEACH" and "Seriously, math would be so much easier if my teacher actually TAUGHT it." Sometimes the bemoaned their fate in all caps messages such as "I HATE THIS FLIPPED CLASSROOM BS" and "LIKE JUST BC IT MAKES YOUR JOB EASIER DOESN'T MEAN IT'S EASIER FOR THE STUDENTS."

Instructors commonly teach online courses without ever having been distance learning students on the other side of the screen. Unlike their interchanges with live undergraduates with whom they can identify as younger versions of themselves, faculty lack a frame of reference to empathize with their students or to understand their frustrations. They also usually know remarkably little about digital interaction design. Ethnographic researchers who work toward developing new software and studying user behavior in technological fields are often first trained to interrogate their own demographic backgrounds,[81] to explore their own online identities and communities,[82] and to report on their own sensory experiences.[83] Yet many who tackle online teaching are discouraged from reflecting upon their own limitations, blind spots, biases, and failures. They assume that they are safe within the panopticon of learning-management systems and marginalize the role of their subjective understanding. In this way the general lack of teacher training in higher education—particularly in comparison to the more rigorous preparation of K–12 instructors—can become more deeply exacerbated by pressures to adopt new and supposedly more efficient instructional technologies, online content-delivery systems, and peer-learning methods in which the instructor wields less direct control. To make matters worse, the instrumentalist assumption that a given technology will only function as a tool to effect the will of a given faculty member ignores the importance of software coding and hardware design and promulgates false confidence about mastery; the active culture of appropriation among contemporary student bodies multiplies opportunities for mutual misunderstandings.

What Does Openness Mean?

In the summer of 2013, the MacArthur Foundation and the MIT Media Lab teamed up to announce the Reclaim Open Learning initiative, which included a hackathon, an innovation contest, and a public symposium. The project was intended to organize more systematic resistance to the "traditional lecture-based, course-based model of campus instruction" replicated by MOOCs and to invest in multiple small-scale pedagogical experiments for "embracing the peer-to-peer connected nature of the web."[84] Writing for the *Huffington Post*, online learning supporter Anya Kamenetz argued that, given how little MOOCs offered in the way of substantive innovation, Reclaim Open Learning deserved widespread public support.

> Even as the "massiveness" of enrollment and "online-ness" of delivery in the MOOC may be new (or somewhat new), in other words, the "course" piece is all too familiar. And what about the Open? Open learning can mean many things to many people. It can mean learning that takes advantage of Creative Commons–licensed open educational resources or OER (of the major MOOC platforms, EDx is open-source, but none have open content).[85]

Kamenetz's comments about the paucity of open access, open source, or open content among corporate MOOC offerings invite further questioning.

It is perhaps useful to think about how the term "open" functions in the rhetorics of the open courseware movement. For example, it is worth asking how open is different from free, in that free universities and open universities are signaled as separate, much as, in labeling software as "free and open source," an important qualification is taking place that recognizes the two terms—open and free—as something other than synonyms. But does the word "open" further compound the linguistic entanglements of the word "free"? As the founding documents of the free software moment indicate, there is a fundamental difference between the freedom of free speech and that of free beer.[86] Just as dissent among those in the digital rights movement may be structured by how sides are taken and how metaphors are chosen in the gratis vs. libre debate, there is a basic difference between an open question and an open meeting. In other words, a course being open for anyone to change and redirect and a course being open for anyone to witness, as long as they are passive spectators in the audience, can connote two very different kinds of educational experiences.

Based just on the number of antonyms that the word might generate, "openness" has far more than two ways it can be translated. What are the rhetorics opposing openness on the educational landscape? Pedagogical models based solely on commercial interest are relatively easy to dismiss as being closed. If academic culture embraces methods that take risks and defy received wisdom and offers opportunities to the previously disenfranchised and disempowered, then an emphasis on security and exclusivity is also problematic for openness. But what about possible virtues—such as stability, selectivity, privacy, and design—that might be opposed to openness? And why "open"? Why not "shared"? How does choosing "open" allow the reinscription of ideologies of ownership and origin?

Moreover, what kinds of origin stories do those seeking to reclaim open learning tell? Linking computer networking to new forms of political economy is certainly nothing new. Beginning in the Clinton era, critics of the divide between haves and have-nots emphasized putting networked computers into classrooms as a means for redistributing cultural capital. Nicholas Negroponte and many others argued that access to distributed computational media would also enhance the face-to-face instructional situation and extend learning experiences by enabling individual exploration of rich, hyperlinked digital collections, thereby fostering educational practices associated with political empowerment, academic research, scholarly citation, collection curation, situated learning, and emergent professionalization. Although the push for the wired classroom continues, there have been a number of critiques of the educational desktop, and then laptop, paradigm. Research on vernacular public-computing practices indicates that several biases may be in play, particularly those that privilege Western ideologies concerning personal computing and consumer electronics, a Protestant ethic that values work and devalues play, and a Romantic notion of childhood as a state separate from a networked adult society. Nonetheless, many who championed putting computers into classrooms began advocating the creation of a full feedback loop, so that classrooms should effectively be put into computers. The advent of the open courseware movement promised to channel the benefits of live teaching to the terminal of a lone computer user remote from a given campus.

In telling the story of open education, former academic and *Hack Education* blogger Audrey Watters has argued that the Khan Academy and other recent start-ups have presented a revisionist history of instructional technology in higher education that erases the work of Canadian educators who created the first massive open online courses.[87] She traces the term MOOC to writing by David Cormier and Bryan Alexander about a course taught by George Siemens and Stephen Downes on "Connectivism and Connective Knowledge"; the course was presented not only to a small cadre of tuition-paying students at the University of Manitoba, but also to 2,300 other students from the general public who, years before Coursera went online, took the online class free of charge. All course content was available through RSS feeds, and learners could participate with their choice of tools: blog posts, Second Life, threaded discussions in Moodle, and synchronous online meetings.

In another chronology of the open courseware movement, David Wiley of Brigham Young University is credited with coining the term "open access" in 1996 in conjunction with his advocacy for expanding educational offerings on the Web and using an "open license" for content other than software. In his course offerings of a decade later, Wiley appropriates several genres championed by literacy experts who specialize in nontraditional learning. For example, a course initially called "Introduction to Open Education" not only used a number of wikis to organize course materials, but also contained game-style "quests" for learners who were organized into different "guilds" and assumed roles such as "artisan," "bard," "merchant," "monk," or "rogue" as they might "level up" in the syllabus.[88]

For those who resist using the canned video presentations, favored by MOOCs, that merely show relatively information-poor talking heads, pedagogical strategies, similar to Wiley's, that are oriented around questing remain promising in new iterations of project-based, problem-based, or inquiry-based learning. For example, in planning to create a multimedia textbook on the history of curiosity, Harvard history professor Shigehisa Kuriyama includes both traditional problem sets and nontraditional games in exercises that foster detective work. He also emphasizes narrative reconstruction activities done by students instead of telling a single, monologic story through traditional lectures delivered at a podium. Using a wide variety of sources that includes images, movies, and texts, Kuriyama's students are tasked with finding connections to spur "rethinking the configuration of the world."[89] Kuriyama observes that humanities educators tend to focus on open-ended thematic questions that give little guidance about how to proceed, rather than the more focused problem-and-solution paradigms typical in language learning programs; therefore, traditional faculty may have difficulty adopting problem solving techniques as they decide how open to make the learning situation.

Because MOOCs are often presented as command-and-control course management systems that promote efficiency in learning by monitoring progress during short chunks of instruction, these curricula generally afford little openness to user-generated remixing for those who want to engage with materials in new ways. Multiple-choice testing throughout the MOOC educational experience further reinforces a sense of having limited options. Some have also argued that this structure makes it difficult to accommodate humanities instruction oriented around the kinds of open-ended questions that Kuriyama describes, although even in computer science courses there may be several correct answers to a given question, depending on the strategy of optimization being pursued. Undertaking different kinds of role-playing activities typifies the pedagogy of Wiley and Kuriyama and has been praised by may educational theorists—whether they urge instructors to adopt the newest technologies[90] or eschew them in favor of "teaching naked."[91] Yet the interaction design of MOOCs focuses on imagining a single standardized user experience with static student roles and almost no possibility of play.

Wiley has complained that many of the salient features of the early Open Educational Resources (or OER) movement have become denigrated by MOOC companies seeking to grab publicity with false claims to radical novelty. He identifies two primary characteristics as particularly critical to truly "open" learning:

> 1. Access to the resource is free and unfettered. That is, the resource can be accessed without the user being required to pay, provide personal information, or jump through any other hoops as a prerequisite to access.
> 2. All users have free 4R permissions with regard to the resource. That is, either by virtue of open licenses or the work being in the public domain, anyone and everyone has the legal permissions necessary to reuse, revise, remix, and redistribute the resource.[92]

When MOOC companies require personal information in exchange for "free" access and claim contractual rights that limit student behavior, they offer an experience of openness that is different from the one envisioned by Wiley and OER early adopters.

Wiley was featured in the collectively authored 2008 tome *Opening Up Education* from MIT Press, which continues a conversation among open courseware stakeholders that was initially sponsored by the Carnegie Foundation. The subtitle of the volume emphasizes how the selected advocacy arguments combine many forms of openness—"open technology," "open content," and "open knowledge"—in arguing for a radical revision of educational practices to suit what the editors characterize as a new class of self-motivated, life-long learners. In the book, Wiley offers what he describes as an "excerpt from a longer chapter" in his "autobiography" written in 2045, in which he claims to chronicle what he calls the "OpenCourseWare Wars" of 2005 to 2012.

Wiley's alternative history of the future (now the past) narrates an elaborate fiction in which he and fair-use law professor Lawrence Lessig must take on Chinese Communists and battle a group of "cyber-terrorists" named the "Libre License League." He even released of video of himself commenting on "deleted scenes" in the narrative.[93] His tale of what was to lie ahead for the open courseware movement, then in its infancy, also includes imagined personal humiliations and intense media coverage of the legalese involved in distance learning, in which Wiley becomes "the stinkiest person in Utah" and Lessig is blamed for "starting the whole mess." At one point, the Hewlett Foundation appears through a *deus ex machina* that brokers a deal between the Free Software Foundation and Creative Commons to end a passionately contested cultural war that is threatening civil society itself. In the "learner utopia" that he imagines, called MetaU.org, most of the academic content is uploaded to the site by students, much as Wiley claims young people do "on YouTube, on Flickr, on Wikipedia, on Facebook."[94]

The potential suppression of rhetorics about paid work and job titles in some digital advocacy arguments is significant, and not all contributors to *Opening Up Education* accept Wiley's summary dismissal of the importance of faculty labor in an open courseware future. For example, Diane Harley reminds readers that faculty, as "the arbiters of pedagogical quality in certificate and degree-granting contexts," are critical to successful open courseware initiatives—"not students, technology professionals, or librarians."[95] Clifford Young similarly elaborates on the importance of distinguishing between "access to education" and "access to information"[96] by pointing out that "educational opportunities" are more than "learning opportunities" because they involve a "sense of agency on the part of the educator" and "some implication of responsibility by a teacher or educational institution."[97]

In addition to asking about academic labor, the role of academic consumption should also be interrogated to understand the discourses of the Open Courseware movement and the cultural imaginaries associated with the online consumer in market capitalism. The fact that universities have become the logo-marked objects of phantom desires associated with lost youth and gained status has not been overlooked by experts in advertising and marketing. For example, Kevin Roberts, of the advertising firm Saatchi and Saatchi, has observed that many universities have attained the exalted status of what he calls "lovemarks" and retain the most sought-after forms of brand loyalty from alumni, who cling to their chosen educational product with strong affection and strong respect.[98] Yet, as William Mitchell asserts his essay "Poison Ivy," the vaunted campuses of the Ivy League often only value the architectures of tradition in order to pander to the nostalgia of an institution's graduates for purposes of maintaining brand appeal in the interest of fundraising.[99] In contrast, the physical space of the MIT campus has no investment in a similar kind of fetishistic appeal, he argues, because it is merely designed to enable certain social, pedagogical, and technological interactions.

Mitchell claims that this is how MIT is fundamentally different from Harvard and other Ivy League institutions, and this might be true of their distance learning offerings as well. It is interesting to observe that MIT's widely heralded OpenCourseWare initiatives present the campus as a mere backdrop to pedagogy lacking in memorable, singular physical locations. In contrast, Harvard responded to the open education challenge by making video-recorded lectures of Michael Sandel's large-enrollment philosophy course in historic Sander's Theater, "Justice," available to its alumni. Thus graduates could relive the course exactly as they remembered it with the sage on the stage.[100] Sandel is perhaps a few decades older than the figure in their nostalgic memories, but he still recites the same lines about Aristotle, Descartes, Kant, and Rawls in a scene of august tradition within an architecture of authority that is over a century old. While virtual Harvard classes with avatars often recreated particular Harvard buildings in Second Life, MIT classes built in virtual worlds were much less concerned with rebuilding virtual landmarks associated with campus theaters of memory.

The launch of the OpenCourseWare Initiative at MIT in 2001–2002 did change perceptions of this highly selective institution, known for its academic rigor and high tuition, as users interacted with its video-recorded lectures and Web-accessible quizzes. The MIT campus also became associated with a number of other kinds of demands for open access, from calls for open academic publishing to open source software in learning systems. Yet not all of these programs were lauded equally by collaborating participants. For example, former MIT professor Henry Jenkins has complained about the inability of media studies to be truly represented in the Open CourseWare curriculum, given administration anxieties about unanticipated breaches of copyright when a film that might be fair use when shown in a lecture hall is duplicated in a digitized form on the Web in direct violation of the initial terms of the Digital Millennium Copyright Act (DMCA) of 1998. Although the DMCA has since been amended to expand fair use for humanities faculty, policy makers tend to be averse to assuming any liability, and MOOC courses have difficulty incorporating multimedia easily.

The open access movement in publishing often focuses on the high purchase prices for print publications that are intended for relatively small, specialized audiences of scholars and the impossibility and immorality of sustaining that economic model, particularly when university libraries can no longer bear the cost.[101] In making the anticopyright, pro-free-culture case, John Willinsky argues for open access to the university's research and scholarship but not necessarily to its classroom experiences. He grounds his philosophical argument for what he calls "the access principle" in concrete narratives that emphasize access to academic literature, particularly for the developing world, as in the case of electronic collections of medical journals essential to the treatment of tropical diseases in Kenya. Willinsky further wishes to make clear that "the human right at issue is not only about enjoying the fruits of scientific progress, whether through new medicines or modified strains of rice. It is also a right to science as a form of knowledge and understanding. It is a right of access to science."[102]

Willinsky's appeal to conscience is aimed at academic audiences who might be otherwise unconscious of their position of privilege, but he stops short of championing access to instruction. He considers the right to read critically and reflectively more fundamental than the right to be part of a specific didactic situation. Thus, open research rather than open teaching is his goal. As Willinsky writes: "A distinction needs to be made here between open access to course syllabi and lectures and open access to scholarly journals and archives. That is, I want to argue, contra Derrida and in the spirit of Kant, that the right to philosophy goes beyond supporting instruction and extending opportunities to enroll in courses."[103] On the level of pragmatic cultural politics, he argues that scholarly foundations are pursuing the path of least resistance by funding innocuous open-courseware initiatives that, by disseminating content with negligible financial value, merely provide feel-good publicity for universities. In his opinion, such open courseware does little to challenge the status quo of the monopolies held by the profitable sectors of academic publishing that control intellectual property regimes—regimes that privilege first-world interests and bankroll many scholarly associations and their professional conferences. He insists that academics are guilty of complicity when professional membership is tied to the asymmetrical subscription model of academic journals that perpetuates imbalances of power.

Those familiar with specific histories of software studies have also asserted that there are significant limitations to the analogies that can be made between open courseware and open source software. Although advocates for professional-amateur learning such as John Seely Brown, who introduces *Opening Up Education*, trumpet claims that the same zeitgeist is expressed in both movements, the authors of "Open CourseWare vs. Open Source Software—A Critical Comparison" disagree:

> Open source is much more than knowledge sharing. It is a system of software development, a legal framework and a philosophy. OpenCourseWare falls short of most of these elements. It does not include the participation of non-MIT staff in the courseware development. Furthermore it is not an integrative legal framework under which other universities might publish their courseware as well, or a philosophy in itself. Although the project's title invites comparisons with open source software, the analogy is probably limited to the concept of making information freely available, hoping for reciprocal gestures from other institutions.[104]

This distinction between culture and content is significant, although those aimed at building community in online, open-access initiatives might claim that they are attempting to create a more robust set of social practices, even if questions of infrastructure and legal codes are inadequately addressed. Manifestos can be written, but traditional disciplines still police boundaries, authorize access to professional privileges, control the production of authorship, and enforce the rules of academic discourse. Preexisting conditions in higher education make digital collaboration extremely difficult to achieve. Furthermore, many competing parties have membership in academia: professors, teaching faculty, graduate students, undergraduates, administrators, alumni, and lifelong learners.

The hierarchy that controls the intellectual property supply chain remains largely intact in the Coursera-style MOOC that depends on exalting authorial fame. Such online courses are sometimes labeled xMOOCs to distinguish them from cMOOCs that apply the connectivist principles of Siemens and Downes.[105] Often xMOOCs emulate the powerful publishing model of textbook companies, even though content is currently free, by offering a corporate slick product that cannot be altered by the educational consumer as the primary vehicle of informational content that shapes curricula. Some argue that xMOOCs might eventually spur innovation in the ossified textbook market as faculty experiment with multimedia content and try to present themselves as lively rhetorical actors on the screen rather than the disembodied affectless authors of static textual knowledge,[106] but others might argue that the iPad rollouts of recent years have shown that multimedia textbooks only standardize curricula further and consign already passive students to be permanently in the audience.

In contrast, as Kamenetz explains, cMOOCs can be extremely polymorphous in modeling different modes of participation; they are generally oriented toward collaborative authoring activities rather than single authorship of a finished product.

> It can mean learning that is self-organized, experimental, peer-to-peer, DIY, badged or otherwise nontraditionally accredited. It takes place where theory meets practice, in communities of practice, in bar camps, hackathons, hacker spaces, Maker Faires, chat rooms, virtual worlds, archaeological digs, libraries, on Twitter, on Vine, on Instructables, on Vimeo, at the after-afterparty to the conference, at the Occupy encampment, in abandoned churches in Pittsburgh, coworking spaces in Nairobi or the Museum of Modern Art in New York City. xMOOCs have never been and will never be the sum total or even the best example of experimentation with truly open learning.[107]

Social, political, professional, and cultural activities may be extremely hybridized. Even when located in established institutions, such as museums and churches, these self-organizing collectives facilitate learning in temporary, autonomous zones that elude formal structures of control.[108]

Within the university, the division of labor among different groups already determines the political, economic, and cultural value of each sector of academic work. As Brown notes, caste systems can be insidious in separating "knowledge workers" from "information workers." In the digital multiversity, research faculty are generally designated as the knowledge workers, while librarians and instructional-technology specialists are categorized as information workers. Collaborative discourse between these groups is hampered by perceived asymmetries in class, and teamwork is stymied by inequalities in rights and privileges. Learning online happens in the context of participation in a complex array of digital media experiences and embodied interactions rather than through an instructional-technology model dictated by scientific management. What is often missing is a media studies approach.

Feminist Challenges

It turned out that I was not the only stealth academic who had signed up for Klemmer's class to learn something about online learning in the Coursera model. A number of prominent feminist scholars of technology affiliated with a very different experiment in open access education had quietly enrolled themselves in the HCI course too. Many of them were suspicious of what they saw as the patriarchal structure of privileging a single source of authority in the classroom, particularly when feminist pedagogies argued for "valuing us all," socially as well as economically, in the spaces of learning.[109] Furthermore, it is worth noting that only about a tenth of the faculty members in the September 2012 Coursera course listings appeared to be female, which would be a shockingly low statistic for any campus diversity officer. Unfortunately, many DIY open education archives are not much better; this is despite the relative importance of feminism and women's studies in the early days of vernacular digital collections for learning that were devoted to making alternative curricula public.[110] For example, in presenting its offerings in the humanities, the Khan Academy notably neglects the study of women, gender, and sexuality and emphasizes a conventional interpretation of cultural production oriented around great books and artistic masterpieces composed by men.

As a recent white paper notes, feminism has a history of open access to education that dates back to the settlement movement work of Victoria Earle Matthews and Jane Addams in the nineteenth century.[111] Feminists tend to describe their classrooms as transgressive,[112] collaborative,[113] engaged,[114] and sensitive to the importance of emotion in learning[115] and interdisciplinary inquiry.[116] This poses an obvious contrast with MOOC teaching premeditated on one-to-many communication from a remote professor focused on single subject matter expertise. As MOOC critic Lisa Cartwright describes it, the feminist distributed-networking project, which answered the trend toward massive courses taught by a few male instructors, brought together working group members from science and technology studies (STS), film and media studies, sci-art, digital humanities, informatics, and critical media practice that also participated in common research ventures.[117] In a September 2012 e-mail to a large listserv of fellow feminists that included Cartwright, Professor Alexandra Juhasz tasked the group with "rethinking and redoing of the unidirectional, massive, somewhat imperialist MOOC."[118] She had been collaborating with Anne Balsamo, Dean of the School of Media Studies at The New School, since 2011, and believed that a critical mass existed among feminist critics of technology to create a vibrant dialogic course that showcased their collective scholarship without reinscribing patriarchal cultural hierarchies that excluded learners.

Even outside of feminist circles, "massive" had become a contested term for many educators.[119] For example, among the antimassive contingent, Berkeley professor Armando Fox has advocated harnessing the computer's capacity for customization to offer a curriculum oriented around the "small private online course," or SPOC,[120] which he later amended to "small personalized online course."[121] Fox, who taught an edX software engineering course, has already announced his intention to teach in a MOOC environment again, but he still questions why massive scale would necessarily enhance learning, especially when large network effects and online depersonalization can magnify the power of "vocal jerks" who attempt to interfere with the intimate individual experiences of committed learners.[122]

As Juhasz was questioning the imperial reach of the massive course, a feminist mailing list was becoming activated as part of a new pedagogical initiative funded with start-up money from the Mellon Foundation; Dialogues on Feminism and Technology included dozens of feminist scholars of technology who would collaborate to teach distributed blended courses on multiple campuses networked by technology in fall 2013. During a year of discussion leading up to the launch of this radically decentered dialogic course, possible acronyms were debated to characterize the work of the group, including MDCLE or "massive distributed collaborative learning environment," MCOLE or "massive collaborative online learning experiment," and finally DOCC or "distributed open collaborative course." The plan for the feminist response to the MOOC phenomenon eventually brought together over a hundred feminist scholars of technology on several continents in FemTechNet. This huge team-teaching experiment, using an archive of shared resources and engaging an extremely diverse population of learners, sought to form ambitious goals for transforming the university. During brainstorming sessions, the feminist group often came up with key terms for the class that were both nouns and verbs—affect, discipline, program, code, play, archive, access, interface, use, brand, remix, labor, and protest—before settling on a stable curriculum that included terms such as systems, infrastructure, and machine.[123]

In thinking about creating the archive of pedagogical materials that the group would share among institutions, Balsamo and Juhasz emphasized the concept of "boundary objects," which had become important in feminist science and technology studies, after Leigh Star published seminal work on how professional and amateur naturalists use common taxidermic specimens. Star described boundary objects as "objects which are both plastic enough to adapt to local needs and constraints of the several parties employing them, yet robust enough to maintain a common identity across sites" that "are weakly structured in common use, and become strongly structured in individual-site use."[124] These abstract or concrete boundary objects also could "have different meanings in different social worlds"

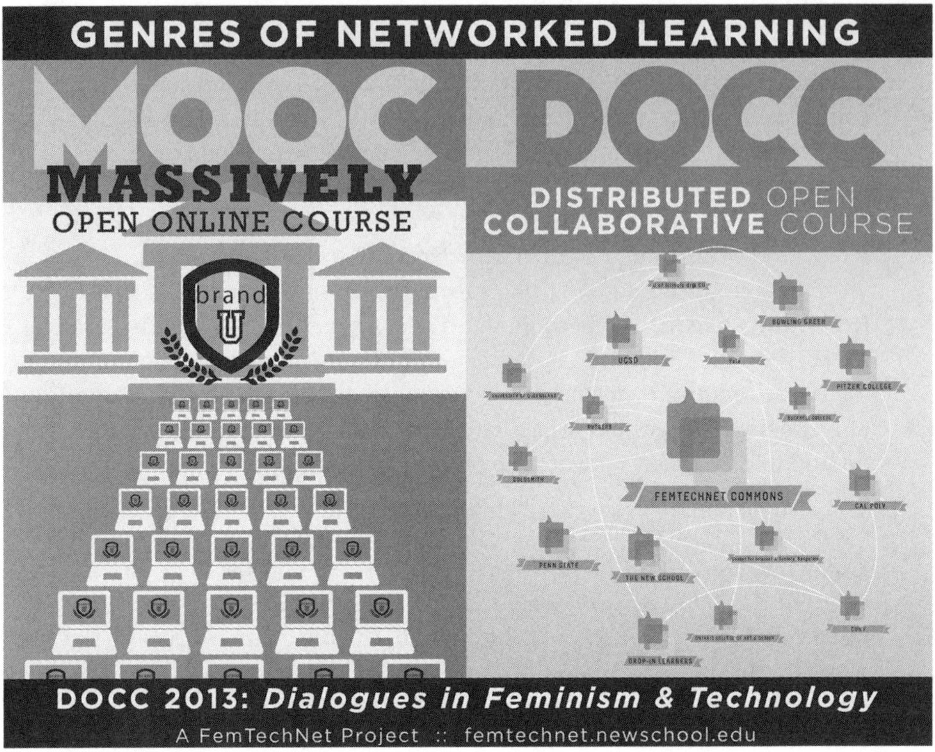

Figure 5.2
Infographic from Dialogues in Feminism and Technology. Courtesy of Anne Balsamo.

despite the fact of sharing a structure "common enough to more than one world to make them recognizable, a means of translation" for "developing and maintaining coherence across intersecting social worlds."[125] Star returned to the idea of boundary objects several times during her career of writing about classification,[126] and the term remained a legacy that the FemTechNet group developed as a way to conceptualize their archive of pedagogical materials, which would be made up of Boundary Objects That Learn (BOTLs). As Balsamo explains, BOTLs are learning materials that are transformed through use when participants annotate materials based on experience and context. This includes readings, media, Web resources, and conversations that have been both submitted to and evaluated for teaching by the network.[127] Although templates were provided for sample BOTLs, users were invited to interact with materials on the terms most appropriate to their own pedagogies and classrooms.

Figure 5.3
Author Elizabeth Losh in a Dialogues on Feminism and Technology sample video. Courtesy of AJ Strout.

Despite its grounding in critical pedagogy or radical pedagogy, the course design placed value on adopting relatively conservative approaches in some fundamental areas. In undertaking this collective initiative, cofounder Balsamo insisted that it was important not to lose the traditional fault tolerance of conventional university instruction, where students could quickly recover from moments of failure in an atmosphere of safety. Balsamo claimed that "failure is an absolutely necessary part of the learning experience" and that the ability to learn from failure—and to help students learn *how* to learn from failure—is a teacher's highest responsibility both online and in the face-to-face learning environment where this large amorphous course would be hybridly piloted.[128] "One of the features that learning online needs to replicate from the traditional classroom is the recognition of the social contract between student and teacher," because students have to trust the teacher to create a space of learning that will be "safe enough for failure." According to Balsamo, students should be comfortable enough to fumble and take risks because classroom errors do not have "life or death consequences." Unfortunately, she complained, "the Internet doesn't forget easily," and momentary lapses of judgment could have life-long consequences for digital youth.

In another nod to the virtues of tradition, cofounder Juhasz reminded instructors that "we already have a model of how people work well with each other: the classroom. We already know how to motivate people to engage there."[129] Although classrooms may have been, at various times, under attack by those seeking to remake the university as a more open place for learning, feminists have used the classroom for consciousness-raising activities for many decades. Juhasz asserted that not only do traditional classrooms benefit from shared experiences of embodiment in a shared physical space and place—they also provided a "structure around which community can form, and where ideas can build, and records are kept." Juhasz insisted that it was important not to lose "institutional support" when jettisoning conventional hierarchy. To adapt the classroom model to a less patriarchal design, she envisioned technology being deployed "to move from an isolated but dynamic space (the classroom) to a networked environment, making the embodied experience of the class and a distributed experience of the class simultaneously possible." This preservative attitude toward the classroom and its moderating functions is often absent in other radically open learning initiatives, where abolishing the paternalism of the teacher and the constraints of the classroom is often celebrated in its most extreme and decentered form.

In their planning conversations, however, the FemTechNet group often struggled to coordinate without a single platform. Even deciding where meetings would take place—on Skype, Google Hangouts, Blackboard Collaborate, Adobe Connect, or an open source platform used by Occupy Wall Street activists—presented challenges to remote pedagogues. In these extended discussions about networked instruction, a number of important issues were aired concerning how such exercises in distributed learning, which were effectively experiments, would be judged by institutions averse to rewarding risk takers or unaccustomed to adapting to all of the ethical issues involved with having students serve as human subjects.

FemTechNet faculty committed to investing labor in the effort; they never viewed instructional technologies as labor-saving devices, and they often expressed skepticism about a sustainable structure for volunteers. At Coursera, rather than finding themselves with the volunteer army of magical helpers promised by cyberutopians, students often felt let down, particularly by their fellow students who shirked duties in peer evaluation. Reading the forum archives about peer grading for my HCI course on Coursera was like browsing the contents of a museum of hurt feelings. Some students felt so abandoned that they sternly scolded peers who had let them down by doing minimal work reviewing projects other than their own. One student wrote that he dug into the server logs and "found the following times the peer graders spent with my prototype (seconds): 44, 60, 115, 1, 63, 63, 25." He lamented that devoting less than a minute would make it "impossible to understand and click over the whole prototype."[130]

Klemmer was obviously sympathetic to such students. At the picnic he talked about using the university's large-format printer to output the specifications of all the student projects so that he could attempt to comprehend them all at once and to appreciate them for the beauty of their multitude and variation. In his final video he praised the ethical responsibility shown by a cadre of students who had self-organized to offer extra critiques to those who had felt neglected in their original, arbitrary peer-review groups. Klemmer told me he was looking forward to trying Coursera again and iterating with new users, but by the end of the picnic he seemed exhausted, worn out from interacting with such a diverse population of students—students who needed so much more than he could possibly give and who, ultimately, wanted his approval as the professor rather than the approval of their peers.

6 Honor Coding: Plagiarism Software and Educational Opportunism

Hello my friends. I am Professor Hans von Puppet. Are you struggling with your schoolwork this semester? Are you canceling a date to take your statistics exam? Well, this is ludicrous! You should hire WeTakeYourClass.com. WeTakeYourClass.com can handle all of your academic needs, whether it be a project, a paper, an exam, or anything for that matter. I'm a professor, but let's get real. The education system is a joke, so don't let the joke be on you. Use WeTakeYourClass.com and get the "A" that you deserve (or you would deserve if you did the work). Anyway they're extremely reliable, professional, and most of all discreet. Visit now and get a free quote.[1]

As the puppet-narrated video for "Pay Someone to Take My Online Class" indicates, a black market of websites exists to subvert the grading standards of distance education and to make a mockery of academic labor regimes. Professor Puppet speaks with a ridiculous German accent and inhabits an office set decorated with diplomas and taxidermy specimens. But the video is not a WeTakeYourClass.com original creative work. The company did not shoot the footage; it was generated from a written script submitted to a site that specializes in generic commercials cranked out with the pompous puppet character. On the Internet, Professor Puppet has been scripted to hawk almost everything, including whiteboard animation, used textbooks, Web hosting, chiropractic services, and electronic cigarettes. Like much of the work done for the company Professor Puppet represents, the costs to clients are assessed by the word.

In prose heavy with irony, the WeTakeYourClass.com website magnanimously describes itself as "a site dedicated to helping students with online classes" and expresses empathy for supposedly disenfranchised students who may "have trouble with numbers" or suffer from other challenging learning difficulties. The actual authors of the work done for credit are described as beneficent "tutors" on the website. The site boasts of "customer service" with "100% secure" results and does not require identifying information for registration.[2] Similar sites, such as Noneedtostudy.com, emphasize that "they do it all," and clients have access to their "academic assistance service" on a twenty-four-hour basis.[3]

Sometimes those who serve as students' surrogates temporarily emerge from their complete anonymity and publish first-person accounts of what it is like to do business for "research services" on the margins of academia.[4] In the *Chronicle of Higher Education*, a writer identified as a "Shadow Scholar" tells college faculty that although they may not know him, "there's a good chance you've read some of my work." Under the pseudonym Ed Dante, this employee of a custom-essay company boasts of completing "countless online courses" after clients provide "passwords and user names" allowing a paid surrogate to "access key documents and online exams."[5] In some instances, the imposter "even contributed to weekly online discussions with other students in the class."[6]

Administrators at academic institutions tend to devote relatively little attention to this kind of wholesale impersonation. They assume that students make alliances with free technologies more often than they do with paid human confederates. Of course, students in online courses also work together without pay in reciprocal peer relationships. Unauthorized collusion has become easier thanks to tools that have become standard for remote collaboration, such as Google Docs and Google Hangouts. As a result, online campuses are exploring new techniques for discouraging cheating, such as identifying distinctive typing patterns or deploying facial-recognition software in conjunction with webcam data.[7] Campus officials are willing to invest in such technology in the hope that new tools can identify and discourage new forms of academic dishonesty, even if new security practices only seem to be necessary because access to computational media through distributed networks has destabilized authorship for many writing in digital environments. By focusing on matching algorithms rather than the social and institutional factors shaping the means of production, technology is able to solve a problem that technology supposedly created.

Plagiarism-detection software, available through the popular Internet-based service Turnitin, is frequently conceptualized as a countermeasure allowing more parity for instructors in a war for control being waged between students and their instructors. Early in the company's history, *USA Today* reported on how "educators are battling back with an arsenal of high-tech countermeasures — anti-plagiarism software, biometrics (thumbprints and retina scanning) to ensure test-taker identity, among others — to help curb academic dishonesty."[8] An informational website for Dallas Baptist University faculty urges professors to "remember that Turnitin is only one tool in your professional arsenal."[9] However, in imagining the equipment of a resourceful enemy force of undergraduates, a Rutgers University business professor complains that "the Internet has just become another weapon in their arsenal."[10] When the company introduced WriteCheck—a new product designed especially for students that would allow them, in advance of submitting their work, to learn how the antiplagiarism algorithm would evaluate that writing—faculty characterized Turnitin personnel as cynical "warlords who are arming both sides in this plagiarism war."[11]

The metaphors used to describe a new technology tell much about how social actors imagine their roles in particular institutional dramas taking place in specific discourse communities. Charles Bazerman has emphasized the importance of such metadiscourse around the location of "knowledge spaces" and the situation of subjects within "symbolic environments" though the maintenance of standardized conventions about originality and imitation.[12] In the current scholarly journalistic debate about plagiarism, these analogies are literalized in ways that have consequences for instructors and administrators who must also provide training in information literacy. Commercially licensed plagiarism-detection software can regulate the boundary between public and private discourse, yet the rhetoric about these services often relegates these discourse practices to mere technological instruments of detection.

In simple terms, an algorithm is a sequence of steps, usually in a computer program, although sequences of encoded procedures don't always require digital technologies to execute. Algorithms can be used for sorting, matching, substituting, aggregating, and many other tasks. In the context of what Tarleton Gillespie has called "algorithmic culture," these operations select the information most important to us and, thereby, often dictate how we participate in public life.[13] As plagiarism-detection software is experienced by users connecting to a common database, such systems also manifest membership in what Manuel Castells has called a "network society,"[14] and thus the algorithm connects discrete nodes representing participants, communities, and texts. Unfortunately, a short-sighted emphasis on acquiring tool literacy from both supporters and opponents of plagiarism-detection software potentially marginalizes the work of librarians, information scientists, intellectual property activists, and others who bring valuable perspectives to institutional choices about the use of surveilling technologies in the writing workshop context.

The Language of Detection

At the University of California, Irvine, where I once served as a writing program administrator, two subsequent chairs of the campus-wide Writing Board, after personal experiences with using the Turnitin service in their own classes, reached entirely different conclusions about plagiarism-detection software. Yet both compared the software to other technologies of detection in making arguments about Turnitin. The first chair, a faculty member from the School of Information and Computer Science, described plagiarism-detection software as being like "airport metal detectors, which we all must endure because we know there are some terrorists out there somewhere."[15] The second chair, a history professor who was also proficient in using electronic media in her teaching, took a dimmer view of the service that the campus licensed to monitor plagiarism. She observed that the software "finds some false positives" and thus creates extra labor for academic subordinates engaged in "looking for false positives."[16]

Both faculty members were early adopters of the technology, but they construed its purposes and functions very differently. Chair One thought that the primary purpose of the technology was to provide security and deter wrongdoing. Chair Two thought that the software served chiefly as a diagnostic tool—albeit a fallible one—that could have little effect after the fact. Chair One located communicative actions in the public space of an airport; Chair Two situated discourse in the private realm of a laboratory or doctor's office. By using the term "we," Chair One indicated that faculty members were also potentially subject to scrutiny; at the same time, he dramatically distanced himself from potential offenders, who were characterized as "terrorists" capable of subverting the safe exchange of persons and products as well as the general good. In contrast, Chair Two referred to an intermediate class of "TAs" who, as graduate teaching assistants, must labor in the place of the professor in doing the work of checking the accuracy of the software's results, and she also acknowledged the implicit inequality of both teacher-pupil and doctor-patient relationships. Unlike the "terrorists" depicted by Chair One, Chair Two focused on possible victims of "false positives." The implication was that like random drug testing, famously lampooned by Barbara Ehrenreich in *Nickel and Dimed,*[17] plagiarism-detection software creates a pseudoscientific bureaucracy that impinges on personal privacy and targets the most vulnerable citizens of a society.

It could be argued that the design of the software itself promulgates attention-getting metaphors about detection. Turnitin, which was developed from 1994 to 1996 by graduate students at the University of California, Berkeley, provides a color-coded "originality report" (from a safe "green" to an alarming "red") for each student paper.[18] Suspect sections of text are highlighted, labeled with a possible source URL or e-mail of a fellow instructor, and assigned a relative percentage according to the total length of the paper. By using an algorithm that looks for matching word strings, the system draws on both Internet sources and a large proprietary database created by the aggregation of individual submissions of student work. With over a million faculty members using the system, and over 20 million students submitting their assigned work,[19] it has quickly become the most profitable and widely used product of its kind on the market. Turnitin is now bundled with a number of course management systems, including Blackboard, and it has sought to expand its operations to peer-review and annotation services.[20]

Use of the Turnitin site is often seen as a quick-fix solution, despite the fact that it still takes a human agent to distinguish between properly cited blocks of text and improperly attributed ones, or if plagiarism has been obfuscated by the use of synonyms or rearrangement of syntactical elements. Just as metal detectors and medical tests must be monitored by live experts, Turnin.com is far from being an automated system, and this argument has been used by both proponents and critics focusing on the role of human digital labor in making final determinations.

The Turnitin company itself seems to prefer the metaphor of the fingerprint match, thereby fostering associations with detective work and law enforcement organizations. The webpage that constituted their vetted "legal document" about privacy and copyright made this analogy explicit.

> To enable a work submitted to Turnitin to be evaluated for originality, the proprietary Turnitin system makes a "fingerprint" of the work by applying mathematical algorithms to its content. The fingerprint is merely a digital code, which relays the unprotectable factual information that certain pre-defined content is present in the work.[21]

A "fingerprint," rather than a signature, is the manual but abstracted trace of the ephemeral physical presence of a unique individual that marks a student's writing. As it is created by the software rather than by the activity of the student, this mark is alienated from the actual written text. Just as a suspect in a police station is fingerprinted, so is an academic essay: "When a paper is submitted to Turnitin, it is fingerprinted using proprietary digital algorithms, and the fingerprint is then compared to the other fingerprints in our database."[22] The students' writing is personified as a potential offender, and access to Turnitin technology by instructors promises to deliver results commensurate to those from an elite national crime lab.

These metaphors from other technologies of detection are important, I will argue, because the technological proficiency associated with academic dishonesty is increasingly becoming associated with analogous technologies. Even the Center for Academic Integrity (CAI), which takes the high ground of traditional rhetoric rather than electronic surveillance, persistently talks about the metaphor of the "radar screen." For example, the CAI website presented "Resources for Getting Academic Integrity on the Radar Screen."[23] In interviews with publications from three separate universities, the center's executive director, Diane M. Waryold, asserts that academic integrity is "on the radar screen" of the center,[24] that good classrooms can get plagiarism "on the radar screen,"[25] and that faculty members are reporting more incidents because "it's on the radar screen."[26] To encourage schools to adopt academic integrity policies, "radar screen" even functions as the label for a developmental stage in the center's taxonomy of how institutions respond to academic dishonesty. The "radar screen" stage is located between a "primitive" stage one and a "mature" stage three.

> Stage two: "Radar screen" This stage describes a school where cheating issues have risen to public debate because of the perceived weakness of academic integrity policies and fundamental concerns with the consistency and fairness of existing practices. Stage two is characterized by early efforts, usually led by administration, to put a policy and procedures into effect, often for fear of litigation.[27]

The radar metaphor is one of combat, specifically of air defense, in which the enemy must first be located in order to be neutralized. A radar-screen level institution has become aware that it is under attack and is thus motivated to overcome its comparative vulnerability in the face of invasion.

All of these metaphors—metal detectors, tests of bodily fluids or tissues, computerized fingerprint matching, and radar screens—point to something other than the production and interpretation of texts in discourse communities. Although they each characterize privacy and individual property differently, I would argue that all these analogies assume an objectified view of communication as product. Either writing is properly a primitive tool to be evaluated by more sophisticated tools, or matter that can be authenticated as acceptable because it is organically appropriate and not alien to the body submitting work. For both metal detectors and drug tests, the aim is to screen out heterogeneous, impure content, which is detected because the presence of a metallic or mechanically complicated component or a contaminated fluid or tissue has been indicated.

Because policing academic dishonesty in cyberspace subverts the collaborative orientation of participants in an idealized classroom, alarmists stage a series of dystopian scenes in which victims and perpetrators are assigned moral positions. In other words, once students and instructors become opponents in a "battle against plagiarism,"[28] most composition narratives take sides. From the position of advocacy for students, we read from Rebecca Moore Howard that fledgling authors have become prisoners who face an "academic death penalty" if they naively commit plagiarism. Moore's language implies that those who are already disadvantaged by class or ethnicity are subject to arbitrary and disproportionate punishment for the combinatory practices of most writers.[29]

In an article about plagiarism-detection software, James Purdy similarly argues that we should be "calling off the hounds," since even antiprosecutorial rhetoric from plagiarism detection software implicitly encourages the hunt for plagiarizers to become bloodsport: the word choice here conjures images of the classic fox hunt—mobs of hunters on horseback galloping after dogs hot on the trail of their next kill. The use of plagiarism detection software, in other words, becomes a hunt for prey. This phrasing also positions the student as a wily and cunning trickster (the folk culture image of the fox) and the instructor as a hunter out for the kill. Moreover, this word choice also suggests an elite class chasing after lower creatures for sport.[30] Although more benevolently described as a "cat and mouse game" in the *Chronicle of Higher Education*,[31] metaphors of the hunt for plagiarizers are not limited to animal prey. For example, one site about appropriate citation practices that advises students in the U.K., where the use of Turnitin plagiarism-detection software has become near universal, compares the nation's centralized academic honesty bureaucracy to a "witch hunt."[32]

If a crime has actually been committed, we debate whether plagiarism is accurately called "theft" or "fraud."[33] If extremely literalistic views of intellectual property are accepted as the norm, faculty members complain of "highway robbery,"[34] and students refuse to engage in peer review practices that might risk having their classmates "steal" their work.[35] According to the *Oxford English Dictionary*, the word plagiarism has an etymology via Martial that includes a prehistory of crimes against persons that include "kidnapping" and "seducing," but by the time we reach the *Webster's Third International Dictionary* the word has become limited to the commission of "literary theft."

Literacies of Tools and Publication

Yet even technocratic metaphors, such as those used about Turnitin, that are not situated within the frame of violence or crime have ideological implications, and not just because discourses of morality inevitably return from the repressed. The American Library Association has characterized this general instrumentalism about new technology as "tool literacy."[36] Unlike higher-order forms of information literacy that are aimed at critical thinking about professional or disciplinary conventions, or about membership in interpretive communities more generally, tool literacy emphasizes competence in the most rudimentary skills of manipulating text (whether it appears in traditional print or as electronic characters and images), often via particular software applications.

The correlation between cut-and-paste replication of materials from the Internet and intellectual dishonesty seems to be remarkably strong. A 2003 study of how Auburn professors defined plagiarism, which indicated little interdepartmental consensus otherwise, showed that the most frequent faculty definition of plagiarism, shared by over 75 percent of the respondents, was "submitting a paper that has been cut and pasted in part or all from a website or papermill."[37] Similarly, in Kelly Ritter's study that formed the basis of her work on the economics of authorship and the rhetoric of paper mills, she discovered that 71 percent of students considered "taking source material from the Internet and using it as my own in a paper or take-home exam" to be "cheating," and only 6 percent thought an "author" could be a person who engaged in "gathering different sources and pasting them together as a collection of writing, then putting your name on that collection."[38] James Purdy has asserted that because technology and writing have become so tightly enmeshed, the technological character of written discourse is in danger of becoming invisible.[39]

However, any convergence of common perceptions shared by faculty and students should be read critically. In 2005 report from the Center for Academic Integrity, it seems that students and instructors have reached different conclusions about the relationship between technological manipulation of texts and academic dishonesty. The Center finds that "most students have concluded that 'cut & paste' plagiarism—using a sentence or two (or more) from different sources on the Internet and weaving this information together into a paper without appropriate citation—is not a serious issue."[40] Moreover, Ritter's study indicated that only 39 percent of students thought that "Writing material for the Internet (either a personal or business Web site)" merited being considered "an author."[41] In Sara Rimer's damning piece in the *New York Times*, she cited research that 38 percent of the undergraduate students surveyed said that they had engaged in one or more instances of "'cut-and-paste' plagiarism involving the Internet, paraphrasing or copying anywhere from a few sentences to a full paragraph from the Web."[42] By virtue of the ease of access, Rimer asserts this conduct was often judged to be "trivial."

Within the community of composition instructors, there has been a contentious debate about the authoritarian dimension of policing the border. This debate centers around the difference between acceptable and unacceptable text practices when "patchwriting" discourses are submitted for academic credit by students who are already culturally disenfranchised, such as those depicted in Mike Rose and Glynda Hull's work on "Rethinking Remediation."[43] Furthermore, particularly when collective and collaborative authorship is so important in the academy and valued in the professional world, as Andrea Lunsford and Lisa Ede point out, to adjudicate student's cases on the basis of potentially selective and arbitrary rules seems manifestly unjust.[44] Moreover, as Lunsford points out in a collaborative hypertext, ownership has been destabilized by the acceptance of postmodern theory in the academy and the advent of new technologies in the larger society.[45] Ironically, college administrators can be some of the worst offenders when it comes to using unattributed prose,[46] and many academic honesty statements used on campus websites and conduct manuals borrow language freely from sources without proper citation.

Currently, the largest and most significant body of published work in composition theory and research emphasizes the risks of excessive legalism in response to the availability of plagiarism detection software. Some, like Rebecca Moore Howard, have situated themselves in strong opposition to stringent administrative regulations concerning academic honesty driven by Turnitin results. She emphasizes the importance of patchwriting strategies across different rhetorical conventions and historical periods: classical and medieval mimesis, works in twentieth-century modernism, postmodern antiauthorial texts, the products of Internet collaborative discourse communities, and—her ostensible main subject—patchwork compositions by unsophisticated student writers who may be intimidated by the demands of institutions of higher education and may respond in ways driven by questions of literacy rather than questions of ethics.[47]

Recently the professional association devoted to campus writing instruction, the Conference on College Composition and Communication, passed a resolution in favor of "sound pedagogical alternatives" to plagiarism detection services on the grounds that requiring students to submit to text matching databases could "compromise academic integrity by potentially undermining students' agency as writers, treating all students as always already plagiarists, creating a hostile learning environment, shifting the responsibility of identifying and interpreting source misuse from teachers to technology, and compelling students to agree to licensing agreements that threaten their privacy and rights to their own intellectual property."[48] However, faculty often feel strong pressure to use the Turnitin system in order to avoid the appearance of laxity, even if they might feel that use of the technology decreases instructional satisfaction for both professors and students.[49]

A number of compositionists taking the non-Turnitin approach have joined together in a multi-institutional research group, The Citation Project, which is studying student papers written from sources and coding each instance of quotation, summary, paraphrase, or patchwriting to analyze how students incorporate the language of others in course assignments.[50] With their findings, these researchers expect to be able to demonstrate that even assiduously cited work might exhibit other problematic features, such as paraphrasing inaccurately or failing to contextualize the opinions of other writers appropriately. In other words, by investing so much pedagogical energy on plagiarism, Citation Project researchers worry that other key issues in writing competency may be overlooked.

Honor code advocates argue that organizing a community of students around administrative regulations will ultimately succeed in producing model students and model citizens. This advocacy position has mostly gained influence in the academy through the endowment of programs and centers. Having looked at the data, I am extremely skeptical of the empirical claims made by honor code advocates, particularly when so much anecdotal evidence indicates that abstract adherence to honor codes has little effect on student behavior.[51] However, beyond this I have two objections to the honor code approach that are grounded in pedagogical best practices: (1) instruction about how to participate in discourse communities is best offered in the context of social practices rather than social principles; and (2) writing is a public act, so having an honest dialogue with students about how texts circulate in the public sphere is more productive than idolizing the stoicism of an honor code dictated only by internalized norms. It is also disturbing that many of the discourses around honor codes involve reactionary and nostalgic rhetorical moves that often gesture toward the authority of traditional military institutions in a politically opportunistic way. Such constructs of integrity are implicitly at odds with diversity and, occasionally, explicitly in opposition to political correctness.[52]

The approach of policy makers who might consider themselves pragmatists accepts some surveillance of textual practices as necessary for understanding the appropriation strategies students pursue, but these middle-of-the-road partial adopters choose to design curricula focused on information literacy rather than academic honesty. This group generally advocates for greater use of and access to plagiarism-detection tools in order to work with students in the context of their actual practices rather than stated ideals. In my own experience as a writing program administrator, I have seen such programs being welcomed by instructors and students, based on surveys of both teachers and learners.[53] This has been particularly true in my own work environment, a public research university with competitive admissions, which maintains a meritocracy with a limited number of spots and a large and diverse population of deserving potential freshmen. After all, some plagiarists have bragged of their exploits in online chats, confident that academic laborers aren't well paid enough to monitor them, which demoralizes everyone. Others have taken the work of their nonnative speaking or working class peers and removed the minor inflections of language or class, and thus have received higher grades on these papers than the original authors. In other words, sometimes avoiding conflict when it comes to the issue of misappropriating texts creates conflicts elsewhere, and approaches intended to avoid alienating some students can actually alienate others even further.

Because Turnitin highlights intertexts, it allows composition researchers to learn much about the patchwriting practices defended by Howard. And because two geographically separated students may buy copies of the same paper from a proprietary paper mill that would otherwise be unavailable to a conventional search engine, Turnitin can also reveal covert textual practices by students who submit whole-text papers from the paper mills, such as those Ritter analyzes. Turnitin members with otherwise disconnected e-mail addresses become associated through their common texts and intertexts. From using Turnitin, we have discovered that there are also many cases where student work is taken from the desktops of roommates or romantic partners without their knowledge and consent, and this practice may involve many kinds of kinship and knowledge relationships.

Academic Labor Not Academic Honesty

It may be much more productive to focus on digital labor, given how ideas about the ownership of academic texts vary across disciplines.[54] I believe that our work in the University of California with Turnitin facilitated campus dialogue about academic honesty, beginning with its implementation in twenty classes in the 1998–1999 academic year.[55] For campus authors, it emphasized a valuable lesson: written work was always potentially public, and texts circulate. Although individualized boutique assignments may seem most desirable for dissuading potential plagiarists, there is merit in having common writing tasks to foster intellectual community and create a basis for empirical composition research.[56]

Unlike an honor code, the 2005 iteration of the University of California, Irvine Turnitin legal release addressed the student in the context of the work experiences of other students. The text of the document read as follows:

Academic Honesty Form for Students

Plagiarists fail the course and have their offense recorded in their School and in the School of Humanities. Violations of academic honesty can affect a student's graduation, financial aid, and eligibility for honors. The University deals with plagiarism cases every quarter, even though most people do not hear about them. No matter how pressured you feel, do not plagiarize; it is not worth it.

...

Please read the following statement carefully and sign it:

"I understand that to protect the value of the independent work that I do in this course, the work of all students in the course may be compared for evidence of plagiarism to the work of other students, both in this course and in others, and to other sources on the Internet and elsewhere; this may involve the storage of students' work on computer systems outside of the university. I understand that my instructor will be the only person who can see my work so stored. I promise to follow all the university, departmental, and course policies about academic honesty."[57]

The consequences that this document enumerates are not part of an eternal, unchanging, traditional matrix of values. Rather, this release represents the rules of a particular institution in a particular time from the perspective of particular disciplines. The emphasis is on external costs, not internal morality: the sanctions affect financial aid, time to graduation, and quality of an awarded degree. The release also doesn't hold students to the lofty goal of "originality"; instead it emphasizes the value of "independent work." The document emphasizes academic labor in a collective academic environment rather than the inspiration of an isolated moral individual governed by the abstractions of higher law. It also puts part of the onus on instructors, who must communicate an institution's expectations clearly, signals empathy to students who are pressured by the conditions for undergraduate labor, and shares information by making public disciplinary sanctions that "most people" wouldn't otherwise know about.

Textual practices around academic honesty are also reinforced by other discursive activities. In collaboration with librarians, writing program administrators have designed information literacy assignments that cover topics such as navigating large databases of research articles, using metadata in library search tools, or understanding how commercial search engines sort results, in addition to more conventional tasks, such as paraphrasing sources for annotations or distinguishing primary sources from secondary ones.[58] In my own experience, these collaborations around information literacy have created opportunities for grant writing, conference presentations, and published work.[59] This is not to say that there isn't hypocrisy in requiring Turnitin (or waxing nostalgic about the days before it was needed), given the university's other positions on issues of privacy and intellectual property. But fostering the cocreation of knowledge with well-designed information literacy curricula showing many kinds of model work could counter the understandable cynicism of those who have studied recent academic scandals in which professors who plagiarize get off much more easily than those who fill their lecture halls.[60]

What Is Digital Property?

On the score of supporting fair use, Turnitin seems to have both a troubling past and a troubling future. Although founder John Barrie has claimed that the University of California, Berkeley avoided purchasing a campus license for the Turnitin product because the university was embarrassed after he pointed out that "cheating was rampant" and thus "the university was dragged through the mud,"[61] the university itself had indicated reservations about the impact of the software on their students' rights to privacy, ownership of intellectual property, and due process, which were detailed in the *Chronicle of Higher Education*.[62] The Berkeley campus also probably felt it had a legitimate gripe with Barrie, because it could argue that the software had been developed by university personnel using university resources. (As a Berkeley doctoral student, Barrie piloted his computer program with students in large enrollment classes on the campus.[63]) Thus, Barrie would have seemed to have capitalized on an investment of public resources and, subsequently, attempted to license the software back to his former employer for profit. Particularly when free and open-source alternatives could be developed—and were at one time being developed at the University of California at Santa Barbara[64]—the near monopoly of Turnitin can be frustrating to administrators desiring more competition in the field.

The company launch of iThenticate in 2004 and CopyGuard in 2005 further eroded the trust of open access advocates, who were concerned about how enhanced copyright enforcement with automated digital software might inhibit the free flow of scholarly information, further constrict how faculty reproduce, attribute, and cite each other's work in the creation of new scholarship, and generate other chilling effects. When CopyGuard partnered with LexisNexis—the electronic legal and public records giant—to "protect intellectual property" with the CopyGuard product "designed to benefit the media and business community," it was clear that both companies planned to optimize access to a "vast collection of content in one place" that would integrate an enormous archive of "6.1 billion searchable documents ... including deep archives" from LexisNexis with the content from iParadigms, the parent company of Turnitin, which was harvested in many years of Web crawling for the plagiarism detection service.[65] The other iParadigms product, iThenticate, would supposedly help "editors, authors and researchers prevent misconduct by comparing manuscripts against its database of over 34 billion web pages and 129 million content items, including 36 million published works from over 465 scholarly publisher participants."[66] Soon the company was issuing press releases boasting of exposing plagiarism across the political spectrum and targeting high profile political pundits, such as the right-wing news commentator Ann Coulter and the liberal columnist Maureen Dowd.[67]

Although far from total information awareness, the potential for surveillance by copyright holders risks the dissemination of information within and between academic communities and the ethical obligations of research universities to provide services for the public good rather than to third-party corporations monetizing faculty content. In a different context, hundreds of professors mobilized for the cause of making digital archives more open after the 2011 federal prosecution of Aaron Swartz for downloading scholarly articles from the JSTOR database resulted in the activist's suicide in 2013.[68] Many researchers posted articles that would otherwise be behind corporate subscription paywalls in a "PDF Tribute" to Swartz.[69] Such outrage is understandable, given the fact that many faculty (myself included) contribute hundreds of hours of unpaid labor to review work for academic journals; then they are asked by the journals' parent corporations to beg their institutions to pay high subscription fees in which a single journal issue generated largely from free labors of love may cost hundreds of dollars to cash-strapped campuses. The Turnitin company may pay lip service to supporting the scholarly mission of disseminating peer-reviewed research by, for example, supporting the travel costs of scholars speaking in favor of plagiarism detection software at conferences of learned societies,[70] but the public relations intentions of such initiatives are clear.

Many in writing programs have begun to consider new allegiances. Too many destructive initiatives against scholarly pursuits have been undertaken in the name of trademark and copyright protections, and writing instructors are often no longer willing to defend the enforcement of intellectual property regimes. Many feel a moral obligation to side with librarians, particularly after library associations took stronger stands on access to information than did associations of research universities, which sometimes allied themselves with the corporate model of financing their operations with patent and product acquisitions. Advocates for understanding appropriation and sharing of digital content as part of broader changes in social norms may no longer comfortably take the side of corporate interests who have claimed an unfair share of the creative commons since the Digital Millennium Copyright Act.[71] Tarleton Gillespie, Chris Kelty, Lawrence Lessig, Siva Vaidhyanathan, and other digital rights activists have been highly critical of this crackdown on peer-to-peer electronic transactions, which they argues stymies creativity and innovation both inside and outside the academy.

However, universities that are fearful of litigation from entertainment conglomerates feel compelled to monitor and limit the capacities of their networks to prohibit the transfer of particular types of digital files. Perhaps nothing dramatized these troubling alliances with corporate interests more than the University of California's widely publicized crackdown on file sharing.[72] The university system was exposed to a significant risk of liability in terms of fending off litigation from both the Recording Industry Association of America and the Motion Picture Association of America, despite the relatively trivial impact of file sharing on the intellectual work done by undergraduates toward a degree. At one time, on some campuses, far more students could be penalized for file sharing audio and video content than were punished for improper replication of text submitted for course credit, even though this form of violation would seem to be much more central to the intellectual and communal enterprise of university life.

As the Turnitin software company aims to expand into new markets, questions about its role in higher education should be asked. Although Turnitin is commonly thought of as a mere vendor by university administrators, it often faces little competition in comparison to other campus contractors. Despite the fact that its algorithms generate important ranking systems by which student work can supposedly be judged more efficiently, not enough attention has been paid to the effects of how such software becomes part of what Bill Marsh calls the "the formalized rule sets" of the university's procedures, in which administrative procedures, institutional definitions, human actors, and different kinds of inscribed texts are part of a Foucauldian mechanism that disciplines all participants.[73] As institutions like the University of Pennsylvania use Turnitin to manage their entire admissions processes, workflow managers may like the fact that applications can be easily vetted for plagiarism,[74] but like any software company its investment in standardization and proprietary modularity may run counter to the interests of its clients. If Rebecca Moore Howard is right, there may even be unintended consequences for access to higher education, if those with the financial resources to buy human "research services," or those with the cultural capital of greater digital literacy, continue to have the upper hand.

Toward an Ethics of Code

There are several ethical issues about the Turnitin software itself that invite exploration from the standpoint of new media theory, although this perspective is usually treated as alien to supposedly nuts-and-bolts policy debates about the problem of student plagiarism and its technological solutions and causes. To break these issues up for purposes of analysis, perhaps it is useful to think about how Turnitin performs many functions as software. Turnitin can be defined as (1) an archive that depends on a database architecture, (2) lines of code created through a collective authorship process, (3) a program that executes, (4) a process that expresses, and (5) a corporate entity that performs rhetorical acts.

As a database, Turnitin may coerce students to contribute to its huge archive of human cultural output earlier every year, now that many primary and secondary schools mandate the Turnitin software. Teachers may unintentionally help the company develop a valuable commodity by enforcing uncritical compliance and requiring the presence of identifying data on submitted documents. Yet, as John Palfrey and Urs Gasser have argued, educators have an obligation to help students think critically about the "digital dossiers" that they create and how the aggregation of data over a lifetime invariably not only represents sacrificing individual privacy but also compromises the agency of those who would wish to curate their own content intentionally, as they might naively understand managing their online identities on social media profiles.[75] I have had students create innovative information visualizations that display their writing histories or online portfolios where they can create archives of written work, but the Turnitin database is one in which they can express little agency. Although changing e-mail addresses and institutions probably clouds the data, instructors should not make the mistake of seeing student content housed in the database as worthless, because it can be of potential value to third parties.

As lines of collaboratively authored code, Turnitin represents the work of many people who write, annotate, debug, purchase, and invariably borrow code. This creates ethical complexity because—as the Critical Code Studies Working Group points out—both software and code "are variously understood as property and as forms of discourse within a community, perspectives which entail sometimes radically divergent ethical positions, complicating the prospect of assigning, or waiving, responsibility."[76] The Critical Code Studies Working Group draws on the work of Italian philosopher Roberto Esposito, who writes that community "isn't a property, nor is it a territory to be separated and defended against those who do not belong to it. Rather, it is a void, a debt, a gift to the other."[77] Esposito argues that the "munus," or gift, that is at the heart of our notions of community and immunity actually functions as a kind of obligation or duty, shedding light on the real ethical stakes in the supposed "gift economies" that operate in code-sharing communities of programmers.

For example, one of the Critical Code Studies Working Group's case studies focused on how the Green Dam Internet-filtering software mandated by the Chinese government took over three thousand lines of code from Cybersitter, an American company that later sued Green Dam for violation of intellectual property rights. Green Dam also lifted considerable amounts of code from open source repositories. Although it may not be possible to examine the Turnitin code for similar evidence of borrowing, such plagiarism-detection algorithms almost certainly don't represent an entirely original work. For example, one disgruntled former Turnitin customer expressed irritation about forcing software engineers to undergo plagiarism checks, given the nature of their collective authoring activities.

> Recall, or realise if you are not a programmer ... code reuse is the holy grail of programming. Programmers should in general, strive to write code that can be reused. That is how programming develops. Large programs are written from smaller programs. The idea is to not reinvent the wheel, and is to write code good enough to serve the same task in other programs.[78]

Thus, the means of production in which the software is created runs counter to the means of production it claims to encourage. Its own code structure is almost certainly appropriated from other sources as well.

As software, the Turnitin algorithm must execute a particular program or, more precisely, execute a series of programs consistently, efficiently, and accurately. Unlike poetry or other written language that can be interpreted or deciphered by humans, computer code gets run and, thus, generates results on machines. However, executing a program successfully without generating errors does not mean the same thing as establishing the truth of a particular condition, such as "X is original" or "Y is the legitimate author of this work." Furthermore, the Turnitin site is careful to specify that it offers no guarantees of being "complete or accurate" and indemnifies itself against the presence of "typographical errors, inaccuracies or other errors, and ... unauthorized additions, deletions and alterations" by claiming "as is" licensing by clients.[79]

Turnitin is not only a cultural product—of the kind that might be displayed in historical museums of the future, in exhibits about the rise of the digital university—but it is also a cultural process that expresses outputs for meaning making. As Noah Wardrip-Fruin argues in *Expressive Processing*, computers can simulate many kinds of devices and generate entire worlds for users to explore. However, many aspects of the design and history of those processes might not be accessible to their audiences, and thus they function like Abelson's famed "ideology machine."[80] Much of this chapter has been devoted to parsing the explicit rhetoric of Turnitin spokespeople, but more study will be needed to make sense of how this common campus software program processes information and thereby promulgates certain values. For example, a clever student may practice tinkering with word strings to lower his or her similarity score and be taught the value of having a particular kind of English vocabulary by interacting with the service. Furthermore, if Turnitin founder Barrie is serious about his "vision" of educational reform that would "exponentially increase classroom feedback" by deploying "efficiency-creating technologies,"[81] it is worth asking how much learners will know about the specific processes directing these feedback loops and which stakeholders will benefit from the efficiencies gained.

The issue of cheating has been a particularly emotional one for faculty members who prize academic honesty and feel that the pursuit of unfair advantages by some students has costs that disenfranchise other students. At the same time, other faculty members argue that patchwriting is a normal part of the developmental processes of student literacy and should not be harshly penalized. Advocates for maintaining a creative commons from which people can borrow in order to create new derivative works assert that Turnitin, the text-matching service that dominates the market, has a corporate model that stifles competition from open source alternatives and exploits intellectual property practices that are antithetical to fair use through its broader business plans. Yet others cannot imagine their campus's academic honesty regime functioning without Turnitin and complain that human beings can't keep up in the Internet era without systems of computing power. As faculty members and students participate in discourses around a particular form of plagiarism detection software, competing positions of pragmatism and skepticism need to be part of larger professional dialogues about privacy, shared property, and academic labor while also promulgating local practices in particular courses that bring these issues to the forefront. More work needs to be done to understand the current rhetorics of plagiarism and to further the education of students about appropriate practices by emphasizing a holistic approach to information literacy that goes beyond training in the use of particular tools. Yet specific technologies do matter when it comes to understanding the effects of software such as Turnitin, and the constraints and affordances of the design of a particular proprietary algorithmic system will inevitably be expressed in the larger culture of higher education.

7 Toy Problems: Education as Product

The camera floats overhead to show a scene of students sauntering around welcome tables on a sunny plaza festooned with balloons. Futuristic music plays throughout this short 2002 historical documentary. Soon the students are filing out of a lecture hall with boxes of new consumer electronics. A voice assures them that they "own this media" and that they should "make it part of your life." Yet the voice also explains that students must sign an agreement and assume "the responsibilities that come with ownership." However, the language of contractual agreements around software acquisition soon changes to utopian rhetoric about the emergent behaviors of the digital generation likely to follow. With their new "mobile computer," these students will "reveal the potential" and reach "insight" about "what is possible." Faculty can only grasp "an inadequate representation of what is possible" having "not grown up in the digital lifestyle."

As the students receive their new Hewlett Packard Jornadas, a student voice bids welcome to the class of 2006. Youthful actors carry on a lively dialogue in the voiceover. A skeptical female voice says, "Yo, what's a PDA. This looks like a palm pilot to me." A male voice praises the new "active campus" system that these lucky students will access with their devices. He moves into rapping about "digital education / it's new tools enhancing communication / you can pinpoint locations / broadcast information / all kinds of information." Students rehearse dance moves, and joyful interactions carry the day.

Mass distributions of handheld digital devices to first-year students on college campuses, such as the one depicted above, offer new ways for undergraduates to both "read" the university and its interfaces and be instructed in new modes of digital decoding by campuses interested in promoting electronic rites of initiation that are markedly different from those of the printed codex. With mobile technologies, students are encouraged to read textbooks, read buildings, read campus infrastructure, and even read each other in ways that seem designed to counter traditionalist anxieties about postliteracy and digital distraction by presenting students as deeply attentive and deeply present in the institutions in which they have gained membership.

How should we look back on a decade of large-scale distribution efforts aimed at giving mobile computing to first-year college students? Four cases studies—the 2002 distribution of HP Jornada Pocket PCs at the University of California, San Diego (UCSD), the 2004 distribution of iPods at Duke, and the 2010 and 2011 distributions of iPads at Seton Hill and the University of Maryland—tell us much about institutional attitudes regarding how students could and should decipher the messages encoded in the current landscape of higher education.

The rhetoric of position papers, historical images, promotional films, and computer platforms that document ten years of campus-sponsored ubiquitous computing exposes the embedded premises in the legacies of what soon may become a forgotten chapter in the history of digital media and learning. By unpacking the basic assumptions of these pedagogical experiments, perhaps we can learn something about our own inheritances rather than assume that the interpretation of electronic interfaces always represents the study of the radically new. Whether we are talking about "clicker" audience-response systems that are designed to promote command-and-control structures in the university or iPhone apps that exploit the mobility and fluidity of new classroom dynamics, attitudes about mobility, ubiquity, space, and place are instantiated in mass-technology distribution efforts. There should be a usable digital past that is archived, curated, and theorized for university stakeholders thinking about these questions for the future.

One of the first mass-distribution efforts, which is rarely mentioned today in discussions about instructional technology, represents an inheritance of my own from Sixth College at UCSD. As the video described documents, in fall of 2002, seven hundred wireless-equipped HP Jornada handheld Pocket PCs were distributed to undergraduate students through the ActiveCampus program. At the time it seemed to make sense for Sixth College to be a participant. It is the youngest of the six residential colleges at UCSD, and it has an innovative culture, art, and technology-themed core curriculum that emphasizes active, hands-on, and engaged learning in its academic plan. Its institutional partners include the multicampus California Institute for Telecommunications and Information Technology. The college also has a general-education digital literacy requirement in programming that entails ongoing collaboration with the Department of Computer Science and Engineering.

The Jornada distribution was meant to dramatize the fact that the campus had developed a mature wireless network by 2002, which provided an infrastructure for information retrieval and location awareness that could be tapped by hundreds of participants simultaneously. To many, this state of wirelessness promoted a cultural imagery of unfettered euphoria, because once-oppressed subjects supposedly had been liberated from locked-down institutions and their constraints. As Adrian Mackenzie has argued in *Wirelessness: Radical Empiricism in Network Cultures*, wirelessness promises "a state of effervescence" that "develops in assemblages of conjunctive relations" and "lies at the fringes of experience, but tinges experience with certain feelings of proximity and attentiveness that may very well not register consciously."[1]

Universities began remapping the terrain of their campuses with digital devices during the late twentieth century with "electronic educational environments" that organized instructional services into navigable spaces accessible through official portals,[2] and they had also introduced "laptop lending" programs to facilitate more use of networked computing resources.[3] But the specific combination of individual private ownership and site-specific public computing that was realized in millennial gadget-distribution programs introduced a number of new features to the topography of the virtual campus. First of all, investment shifted away from campus

Figure 7.1
Students with HP Jornadas at the Sixth College Explorientation in 2002. Courtesy of Ioana Patringenaru.

computer labs, digital playrooms, smart classrooms, or other confined spaces equipped with fixed workstations; in addition, the labor of maintaining and updating devices was transferred directly to the students themselves. Second, student usage of signature electronic products in a variety of settings indicated that the interface for a particular computing behavior could be considered a marker of status, a sign of membership, a token of belonging, and a badge of authority. Third, gadget distribution changed the meaning of distance between campus locales, but it did not generate a radically new state of omnipresence, since students still largely traveled by foot from point A to point B, even if they might be doing real-time voiceless coordination between two spots. (If anything, it might have made the lag between time of communication and time of contact more noticeable, so that time may have seemed to have been slowed down rather than speeded up by access to the machine.) Fourth, the devices disciplined the student body in new ways, as learners were kept within particular perimeters of coverage during activities, where they also acquired the personal digital assistant (PDA) postures needed to read specialized screens.[4]

Location awareness created a number of other ancillary effects, particularly when single operators could be associated with specific devices. Since signals sent from Jornadas were to be tagged with students' names, freshmen could effectively become subject to roll call at any time and place, and with the ActiveCampus system they could be marked present (or absent) even during off hours when they had not volunteered any attendance information. Scholars studying newer services such as Foursquare, Google Latitude, and Facebook Places have noticed similar deanonymizing effects. Although today's undergraduates might appreciate the convenience of integrating Internet access with daily life and the fact that "the web instills locations with data resources, making those physical locations part of the web,"[5] their trajectories of seemingly autonomous movement across different infoscapes can now be datamapped intensively and, potentially, could be capitalized on or manipulated. Certainly, now that third-party applications on students' smart phones have become the norm, regulations governing student privacy rules have not kept up with the patterns of campus use; it would be difficult for universities to regulate campus data airspace effectively at this point.

Surveillance and Sousveillance

The planners of the ActiveCampus initiative wanted to make certain that the distribution of the HP devices would provide much more than a transient product-placement opportunity. So, to imbue the event with meaning, they enlisted the media artist Adriene Jenik to serve as art director and design rich user experiences for students equipped with the devices. She created treasure hunt activities and other games in which virtual objects could be gathered, such as "Mystery History," "Maprobatics," and "Finders Keepers." Students were expected to complete tasks that ranged from decrypting the biographies of important UCSD alumni to finding their way through labyrinthine buildings in search of prize coupons. Students could also participate in activities of social surveillance by using a buddy system to identify the presence and location of members organized under "Actors and Groups."

Jenik was inspired by her experiences to develop the dystopian rather than utopian possibilities of such devices, and she soon was collaborating with Ricardo Dominguez, Michael Trigilio, Natalie Jeremijenko, and other media artists affiliated with UCSD—along with science fiction author Kim Stanley Robinson—to create SpecFlic, a series of public installations about alienation from traditional print artifacts. Jenik launched what she called a "distributed social cinema project" constructed with layers of media forms. The narrative was set in a library in 2030 that served as a node in a universal knowledge repository known as the InfoSphere. The polymorphous InfoSpherian/librarian who served as guide and muse on the giant hyperwall display screen mediated her own face with interfaces as she donned various veils, glasses, and cosmetic masks. Along with members of the general public, a number of Sixth College students enrolled in Culture, Art, and Technology's collaborative, project-based media production classes also contributed footage to the archive that supposedly represented the 2030 datastream. Thus the feed of the Infospherian masters appeared to be mixed with subversive user-generated content. Using Steve Mann's idea of "sousveillance," in which surveilled political subjects can express resistance by gathering information with wearable or portable devices,[6] Jenik enabled a "Sousveillance Grid" that encouraged "bystanders to assist in documenting their surroundings."[7]

Jenik was not completely incorrect in seeing how distributed computational platforms could easily serve command-and-control agendas. One of the most popular instructional uses of the HP devices by faculty was to support periodic spontaneous quizzes, so that students would be held accountable for their presence in large lecture-hall classes. A similar use continues to this day; computer science faculty still use clicker response systems in courses serving Sixth College students. As ActiveClass designers provided a mechanism for polling students continuously, they acknowledged that

Figure 7.2
2005 SpecFlic installation. Courtesy of Adriene Jenik.

power was being exercised "politically," although they imagined this politics as participatory rather than hegemonic, and they separated out the relationship of strategy and tactics played out in the spatial architecture of the room from recognizably political dynamics, so that "what we have termed the political aspects (i.e., the relations between professor and students)" is separable from "the physical aspects (the desks, artifacts, and layout of the classroom)."[8]

In recent years, incorporating sousveillance activities into the pedagogy of credit-bearing courses, as Jenik did, has become much more common. For example, University of Maryland professor Jason Farman assigned a mapping exercise that had students document the presence of security cameras across campus. As Farman describes in *Mobile Interface Theory*, students realized in the course of doing the assignment that the cameras weren't actually designed to make people more secure; instead, they were intended to protect university assets and infrastructure, so that "the embodied relationship to this space emerges … as always in relationship to commodities."[9]

While affiliated with the ActiveCampus project, Jenik found herself contemplating the vulnerabilities of what it meant to be "off the grid." As she noted, the "UCSD campus wireless network, though extensive, is not exhaustive."[10] She suggested that selecting "non-wired areas as a footprint for investigation" could enable people's ideas about "community, technology, and technology-enabled communities" to be assessed critically. Under Jenik's guidance, a series of gamelike activities would help students become aware of an important concept in art and graphic design—negative space—in order for them to rethink the figure-ground relationships of the campus. Given that one of the purposes of the project was to rejoice in the campus's supposed great leap forward in ubiquitous access to the Internet, Jenik's decision to landmark campus cold spots rather than hot spots might have quickly marked her artistic interventions as counter to the dominant positivism of the project.

Observing Infrastructure

According to computer scientist Bill Griswold, one of the goals of distributing such devices was to allow students to see and be seen in new ways. The Jornadas were intended to display information that would otherwise be occluded. Griswold argued that his own institution needed this enhanced display mechanism, because it was difficult for students to comprehend the many activities of the university by merely relying on what could be seen without ubiquitous computing technologies—through windows, on websites, and on informational monitors.

> Since a campus institution is typically a physically aggregated entity, displaying an institution in a transparent form and showing its mediated sources of learning "inside" it (or even next to it) is a natural way to convey mediating relationships. Depending on the possible relationships between the learner and the learning source (including role reversal), participants may need the ability to talk—as well as see—through walls.[11]

Like a kind of X-ray device, the HP Jornada computers were supposed to allow students to look through the walls of the university buildings and into the secret departments, laboratories, research units, and other organizations housed inside that they would otherwise never see.

According to press releases from the period, Griswold was a serious believer in what was described as "transparent infrastructure," which one writer joked was much like the futuristic "transparent aluminum" of *Star Trek*.[12] Of course, as a rule, infrastructure tends to be hidden from view. As Genevieve Bell and Paul Dourish observe, "the infrastructures of daily life—the electricity system, the water system, telephony, digital networking, or the rest" require someone to "lift the cover, peer behind the panels, or look underneath the floor" to make visible the "mess" that is "never far away," which might consist of "a maze of cables, connectors, and infrastructural components, clips, clamps, and duct tape."[13]

In many ways, the ActiveCampus project performed what Geof Bowker and Leigh Star have called an "infrastructural inversion" that can express resistance to the fact that infrastructure disappears unless the need for its presence is made manifest through its absence. In other words, infrastructure goes unnoticed unless it is broken down. By "recognizing the depths of interdependence of technical networks and standards" and "the real work of politics and knowledge production,"[14] infrastructure can become an object of study, according to Bowker and Star, and it becomes possible to see unseen support systems, such as those that provide the basic framework of a postsecondary institution.

But how precisely should infrastructure be defined? Star and Karen Ruhleder famously define it as follows: "It is both engine and barrier for change; both customizable and rigid; both inside and outside organizational practices. It is product and process."[15] Following Star's reasoning, infrastructure can be both abstract (such as common standards, classification systems, and schedules) and concrete (such as cables, antennae, and maintenance tunnels). It can be composed of inanimate material objects (such as cooling systems) and human actors (such as staff hired for monitoring equipment). In addition to being a "what," infrastructure can be a "how" and a "who"—and even a "when." As Star explains, something "becomes infrastructure in relation to organized practices," so that—within a given cultural context—timing dictates the definition of infrastructure. In this way, "the cook considers the water system a piece of working infrastructure integral to making dinner; for the city planner, it becomes a variable in a complex equation."[16] However, Star also notes that the inevitable "tension between local, customized, intimate and flexible use on the one hand, and the need for standards and continuity" can become exacerbated by "the rise of decentralized technologies used across wide geographical distance."[17] Unlike distance-learning schemes spanning continents, the hybrid-learning ActiveCampus system focused on a narrow region. However, it still had to incorporate technologies and services from multiple corporate sponsors, including HP, Microsoft, and Intel, as students connected to local infrastructures.

As a convert to Star's type of infrastructural analysis, Philip Agre does not share the enthusiasm of many of his fellow computer scientists for distance learning. In fact, Agre expresses considerable skepticism about "a certain simple story" in which "classes will be conducted over the Internet," "students will pick and choose the classes that best suit them," and "the resulting competition will improve the prevailing quality of instruction."[18] Working back and forth between "network architecture and the technology and sociology of network applications," he considers the university to be an elaborate assemblage that encompasses its systems of governance and "many activities ... that do require physical proximity, such as theater and dance, laboratory classes requiring access to expensive equipment, athletics, and socializing." These local nodes of situated interaction were also the kinds of hidden assets on a campus that ActiveClass was intended to uncover.

In his essay on university infrastructure and institutional transformation, Agre argues that two critical but low-status aspects of a campus are too often overlooked: (1) "the library, instructional networking and computing, the telephone system, media services, the campus bookstore and course reader service, the course catalog and schedule and many other paperwork-handling offices"; and (2) the fragmented and informal instruction offered sporadically in "teamwork, consensus-building, professional networking, library searching, event organizing, online conferencing, basic Web site construction, brainstorming and innovation processes, study skills, citation skills, and so on" that is urgently needed for personal and professional development.[19] In making visible and then reorganizing these two essential dimensions of the university for purposes of networked teaching and learning, he argues against the false efficiencies of standardization that might push distance learning as a way to abstract the messiness of existing infrastructure.

In thinking about the question of infrastructure in relationship to complex collaborative activities of knowledge making, it is worth considering a different paradigm for deploying ubiquitous computing from the ActiveCampus model: the "interactive workspace" or "iRoom." In these specialized and resource-intensive university workspaces, multiple devices can be coordinated and displayed with shared screens and shared input devices, such as smart boards that can convert handwriting into digital text. The work of Terry Winograd and other Stanford researchers has been central to the development of technologically enhanced classrooms following this model. Although Winograd's research group imagined devices that could be decoupled from specific sites, programs, and phenomena, they believed that fixed locations would make coordination between users' personal devices much simpler and more meaningful. Part of the group's logic in adopting this approach had to do with the issue of infrastructure dependency, as well as the need for access to more powerful and programmable computers than PDAs and, later, tablets could provide.

A large body of work goes by the names ubiquitous, pervasive, and mobile computing. Although many people use the terms loosely and often interchangeably, we find it useful to categorize each contribution according to the degree to which it is infrastructure-centric. By this we mean that a ubiquitous-computing device such as a PDA can make strong assumptions about its environment: the availability and quality of network connectivity, and the availability and quality of computation embedded in the infrastructure to assist PDAs in interacting with the rest of the environment and even with each other.[20]

In Winograd's own infrastructural inversion, the PDA serves as a networked actor that makes "strong assumptions about its environment," particularly about its infrastructure, and thus must be situated in areas with stable support. This kind of object-oriented ontology, which focuses on the smart device rather than its operator, can reveal networked relations that might otherwise be unseen.

As more students, faculty, and staff come to own personal smart phones with individual data plans, network administrators are complaining about how these devices may become much more infrastructure-centric than they initially seem. Specifically, budget-conscious smart phone owners often tap into the university's wireless services rather than pay per-megabyte mobile broadband costs for these devices. This creates unprecedented demands on networked computing as staff members catch up on TV shows during lunch breaks, or as students share videos over the weekend. However, attempting to police these activities may prove to be as fruitless as earlier crackdowns on campus file sharing. Besides, the university so often benefits from the blurring of work and leisure, and the emotional investment of participants in campus life, that to enforce barriers suddenly would be hypocritical, especially given how much university work is done off-hours and off-site. Much like the always-on professionals of today's digital labor force, described by Melissa Gregg in *Work's Intimacy*,[21] the on-call availability of faculty and students on smart phones strengthens the extractive capacities of the university as an institution.

Facing Interfaces

Theories acknowledging the promise and limitation of the university's infrastructure have become important in a number of recent instructional-technology initiatives, as the work of Griswold, Agre, and Winograd demonstrates. Device distribution programs are also important in promoting a culture around new interfaces on campus. Such interfaces both "screen" and "screen out" information, as Lev Manovich has argued, which can present challenges when orchestrating complex student interactions with a university campus.[22]

In addition to creating an architecture with the transparency only fitting for a public institution, Griswold's group apparently also hoped to provide a greater variety of configurations of membership for students desirous of belonging to this extremely heterogeneous organization. In studies published on the ActiveClass experiment, concepts of space and place are used as the stages in which these negotiations of aggregated social actors can take place.

> A department is not just an aggregation of interest, but is a full-blown institution providing services for its aggregate of people, including working spaces, meeting spaces, seminars, opportunities for chance interaction, equipment, curricula, degree programs, funding, etc., to enable and encourage the processes of learning.[23]

Using the imagery of floor plans and cutaway views, Griswold's rhetoric abstracts the taxonomy of campus meeting spaces and maps it onto the flowcharts of capitalization and progress toward degree, thus encouraging an opportunistic approach to the literal seats of power in a college that might allow students to finally have a meaningful place at the table with decision makers.

Griswold recognizes, however, that any technologically and culturally complex multiversity has regions with distinctive configurations of features that might resist mapping in a single dimension. As Griswold remarks, "UCSD is divided into residential college neighborhoods. Each department sits in a college neighborhood and is nuanced by it, but does not belong to that college; it belongs to a school. Each faculty member belongs to a college, however, and of course a department."[24] With ActiveClass, Griswold aspires to facilitate interdisciplinary communication by rethinking the redlining of particular academic neighborhoods.

Unlike those who imagine a confident and technically fluent digital generation, Griswold's group fears that potential rootlessness away from home could cripple students and deprive them of any agency in this new environment.

> With the arrival of the baby boomers' children, the University of California, San Diego (UCSD) is quickly growing from an intimate small town into a bustling city full of unfamiliar faces. ... This rapid growth has brought numerous "big city" stresses. ... Unfamiliar faces are everywhere, even obscuring those that you know.[25]

ActiveCampus researchers posit that these hand-held devices might guide students' participation in a campus that may be less like the geographies of their intimate cul-de-sac communities of origin and more like those of the alienating "unfamiliar faces" of contemporary networked publics, where students must negotiate overlapping opportunities and occasions without relying on their high school friend networks. Although Sherry Turkle has famously argued that portable devices only cause those with private computing in public contexts to become more deeply "alone together"[26] and that computing exerts a "holding power"[27] that can intensify social isolation, Griswold's group predicts a scenario in which social media gadgets will plug into a much-needed platform for emotional association and collective problem solving.[28] According to this argument, the interface of the ActiveClass system should have only a positive mediating influence that facilitates connection, rather than one that blocks person-to-person interaction.

In contrast, Alexander Galloway describes a very different kind of "interface effect" by "showing how digital aesthetics both facilitate and prohibit political encounters" and describing how a range of devices might produce "an autonomous zone of interaction orthogonal to the human sensorium, concerned as much with unworkability and obfuscation as with connectivity and transparency."[29] Galloway also insists that far too much attention has been paid to the screens of ubiquitous computing devices, so that it has become difficult for critics to meaningfully engage with the effects of nonoptical interfaces—such as sensors, keyboards, and mouses—in addition to the interface effects of computer memory, algorithms, and protocols. For Galloway, an interface is always less a *thing* than an *effect*, one that is associated with practices and dynamics rather than with static, designed objects that are passive artifacts to be distributed and commodified.

Griswold does acknowledge the possibility that access to hand-held interfaces may work against the pedagogical aims of faculty, but he tends to focus on individual ownership of cell phones as the most salient difficulty obstructing full classroom participation.

> Because these institutions operate through proximity, they function less well when people are not "there" on a full-time or full-attention basis. ... When such obfuscation is combined with a busy schedule, conflicting priorities, distractions, interruptions (half of UCSD undergraduates possess cell phones), it is not surprising that many opportunities are missed.[30]

Although the classroom may be punctuated with interruptions from one device, using the ActiveClass technological suite with another device makes transient and anonymous social actors become visible and activatable in more productive ways, so that the quality and quantity of class participation can be optimized.

As smart phones entered the campus landscape, and as platform competition among palm devices changed the dynamics of the consumer electronics market, the Jornada was jettisoned as a classroom tool. Rather than serving as a revelatory interface to focus attention on the articulated components of the knowledge structures of the university, smart devices were often perceived as causing scattered attention; this was especially true as the distraction panic adopted a bigger profile in the larger cultural conversation about digital media and learning and intensified conflicts between faculty and students about the battle between "our" technologies and "their" technologies.

Many instructors considered the Jornada to be a problem rather than a solution for their multitasking students. They claimed that it complicated classroom management without any obvious benefits. Ironically, some faculty members working with the Jornada—including those who tolerated digital distractions—reported that user difficulty in operating the HP tool could be cited as a virtue, because students could not multitask during class.

> Finally, the last essential element for the fitness of ActiveClass was the professor's tolerance for using PDAs for "unapproved" activities such as instant messaging and playing games. Both of our professors took the view that it was their responsibility to create an environment that attracted the students' attention, and thus tolerated such activities as long as they didn't distract other students in the class. In this case, the small display and pen-based input—cited as problems in the next section—were a benefit, as they induced minimal distraction.[31]

These design defects inhibited fluent digital behaviors and interrupted the interruptions of other devices. Thus students were forced into deeper concentration as they monitored input and output more closely.

Despite the fanfare of the launch and the hoopla surrounding the testing phase, students never really took to using the device. Researchers noted that some students even asked them directly, "How can I use this to study?"[32] Within a year, even the early adopters had abandoned the ActiveClass platform, and the devices were no longer required to complete assigned class work. Much of the distributed equipment appeared on eBay; many of the Jornadas were relegated to e-waste. But gadget distributions continued.

The Decade of Gadget Distributions That Followed

In 2004, as part of a university initiative to encourage creative uses of technology in campus classrooms and dormitories, Duke University distributed 20GB Apple iPods with Belkin voice recorders to over 1,600 entering students. At least fifteen courses incorporated the iPods into instruction involving 628 of the students. The gadgets were used for recording lectures; capturing sound from fieldwork visits; skill and drill work in music, media, or foreign language classes; and file storage and transfer. UCSD PhD candidate Tara Zepel recalls being a sophomore during that time and her initial envious feelings toward the freshmen who had each received a customized iPod with a "cool silver backing complete with an engraved Duke shield." Later, when she realized that the iPods were "part of an educational experiment" with the Apple Digital Campuses program, she said her "envy turned to ennui" as the news media "raged on about the viability of using the music-playing device in the classroom." Eventually she felt embarrassed about her preliminary reactions. After all, she preferred live performances to recorded tracks. In her junior year she "took one of the classes that warranted the coveted device" on the history of radio production, where students were expected to listen to old shows and create their own updated versions. She used it a few more times to listen to lecture podcasts and to record an interview, but she did not see it as a transformative technology.[33]

Although Duke committed to the proprietary software because the associated user interface was already attractively familiar to many, administrators ultimately chose not to continue the program beyond the pilot year. Faculty described exasperation with the many challenges of integrating multiple systems for content storage, access, sharing, and distribution without full campus tech support, and students often found the devices most useful as hard drives. Furthermore, there was no system for bulk purchasing or licensing of content for academic use. Curricular planners noted the iPod's inherent limitations as a functioning input device, particularly for working with text, as a significant drawback for creating student user-generated content. Because faculty and students lacked awareness about the device's possible uses, a lack of training resources contributed to its unsustainability.

Nonetheless, the legacy of the iPod distribution at Duke included several positive effects. A report on the project indicated that the "unexpected publicity" for the initiative generated by "the cultural phenomenon of iPod use" resulted in "hundreds of unsolicited inquiries and invitations for collaborations" that solidified Duke's reputation as a leader in the digital humanities, cultural informatics, and advanced technologically enhanced pedagogy.[34] Despite the public relations benefits from this newsworthy instructional-technology event, participating faculty often resented being stymied by the copyright restrictions of iTunes and objected to the locked-down Apple operating system that prevented creative appropriation. Soon the novelty wore off for all parties. The experiment was not repeated the following year, although academic computing personnel claimed that they had gained lasting benefits from the launch by getting faculty comfortable with sharing pedagogical ideas across disciplinary silos and collaborating with instructional technologists.

Currently, iPad distributions are the rhetorical occasion capturing headlines, although campuses often report mixed results.[35] Despite the fact that the adoption of e-books at one time promised to break textbook monopolies protecting high price points, the shift to a license economy from a property economy also has had its perils, so that students who purchase content for electronic devices may not necessarily own it in the same way that they owned, and could even recover investments on, more traditional curricular materials. Furthermore, the ways that class, race, and gender operate for the purpose of marketing efforts aimed at perpetuating the prestige of the Apple brand may be different from the ways that we would wish diversity would operate in the consciousness-raising activities of the university, where questions of difference might be understood through the lenses of multiple forms of intersectionality, rather than received with the flattened affect of narrowcasting and niche marketing. Thinking more globally, it may also be valuable to be skeptical of gadget distribution projects in general—such as the struggling One Laptop Per Child program—in applying a missionary mentality that privileges placing value on individual Westernized consumer commodities over indigenous social practices oriented toward the sharing of digital resources.

The distribution of iPads to all faculty and full-time students at Seton Hill beginning in 2010—with iPad minis added in 2013—was an important institutional marketing opportunity for the small Pennsylvania university, which is often confused with the larger Seton Hall in New Jersey. With catchphrases like "An iPad for Everyone" and slogans like "Think Outside the Classroom," the initiative telegraphed ambition in every aspect of its verbal and visual rhetoric. Its launch event celebrated a ritual "death of the textbook," featured the "iPad dancing troupe," and placed an impersonator posing as Apple founder Steve Jobs center stage.[36] Unlike the more guarded Duke report, which noted the burdens on staff created by publicity, the Seton Hill report from 2010 to 2011 enthused about the boon for public relations and boasted of mentions on television shows such as *Good Morning America* and *Jimmy Fallon Live*.[37] The report also bragged about the speed and coverage of its Enterasys wireless network and the campus's history of "infrastructure enhancements." After all, like the HP Jornada, the Apple iPad was an infrastructure-centric device.

Because the institution was generous with faculty training programs and took an inclusive approach to distribution that even reached out to adjunct instructors, the Seton Hill program was considered to be an exemplary model by many instructional-technology experts.[38] Faculty enthusiasts included English professor Dennis Jerz, who argued that iPads in the classroom could facilitate better feedback on reader reception through the analysis of student annotations that were now public rather than private on the page.

> I love the idea of being able to see the marginalia that students post in their texts. … I'd love it if I could pick a random chapter of *The Scarlet Letter* and see what a student has highlighted. (I do something similar, of course, by asking the students to post a quotation and a brief statement about each assigned reading, but if I could see the students' private annotations, I'm sure I'd learn a lot more from their passive, ambient annotations, rather than the piece they choose to display to the public.)[39]

Jerz continued to be an enthusiast, and in 2012 he created his own iPad app for teaching and learning the sonnets of William Shakespeare.

Although the University of Maryland also had its own ambitious iPad distribution in 2011, faculty and administrators there have been much less enthusiastic about continuation, and it is likely that the program will be phased out soon. During the implementation phase, Professor Hasan Elahi asserted that the portability, screen size, and page dimensions complemented existing social computing activities oriented around recognition and rites of membership. Elahi also noted that the iPad was particularly well suited for locative freshman-orientation exercises that might involve Quick Response (QR) codes or group programming lessons. "Icebreakers" to encourage Digital Cultures and Creativity (DCC) students to get acquainted ranged from doing the hokey pokey to swapping university program-themed t-shirts.

> In an icebreaker twist, each DCC student received another classmates' T-shirt. Students were asked to use their iPad to figure out whose T-shirt they had actually been given. Each T-shirt was personalized, not with the student's name, but with a unique QR code that was ironed on. … For every student, a webpage was created displaying their name, major, and hometown in an image that linked to that student's online presence. This intentional mix-up on our part allowed the students to familiarize themselves with their iPads and also provided the opportunity to meet one another.[40]

Elahi's interest in fostering social surveillance in the incoming first-year group was not without its ironies, since—like Jenik—as a media artist he is known for doing work about surveillance and sousveillance.

Other Maryland faculty expressed deeper reservations. American Studies professor Farman expressed disappointment at not having more versatile iPhones with which his class could work— iPhones that would also have been more appropriate for his scholarly research on the culture of mobile telephony. Nonetheless, he appreciated how the iPad had become important as a stable platform for the development of electronic literature, and he even worked with his students to create a piece for the e-lit exhibit at the Modern Language Association. With his small class of seventeen students, he was also able to use the iPads for "a Twitter backchannel, site-specific quizzes, participatory surveillance, location-based gaming, and locative storytelling projects."[41] At the end of his iPad trial period (winter of 2012), Farman expressed regret at no longer having access to the devices as teaching tools; however, by 2013 he supported the campus decision to decommission the program on the grounds that it offered too little bang for the buck.

At UCSD's Sixth College, we are currently experimenting with other kinds of ubiquitous computing technologies to exploit new possibilities for aesthetic, political, social, and critical engagement that could avoid the painful aspects of gadget-distribution initiatives. Pedagogies around appropriation and media archeology, group copresence around shared deliberative spaces, and low-tech media that include print artifacts, clothing, and convention swag offer new ways to explore questions of community, infrastructure, and mediation. After all, in thinking about the augmented reality that adding new interfaces to the classroom can provide, even e-mail or text messages can serve as a kind of additional layer to sustain a platform for a meaningfully augmented reality if instructors have given students access to electronic communication channels associated with their personal affect and identity.

In *The Virtual Window*, Anne Friedberg argued that most interfaces tend to have long histories that connect theory to practice and project certain kinds of cultural imaginaries on their surfaces. For example, in the case of Microsoft's Windows technology, applicable knowledge from the society of the screen might range from the work of Leon Battista Alberti to that of Paul Virilio.[42] If mass distributions of handheld digital devices to first-year students on college campuses offer new ways for undergraduates to read the university and its interfaces, perhaps we should also be encouraged to read the university differently ourselves. In particular, we could consider questions of face in connection with interfaces in ways that respect the ecologies of rapidly evolving reputation systems in which students and instructors are both coconstructing knowledge and attempting to regulate each others' conduct.

The ActiveCampus "Explorientation" imagined college as a complex and opaque black-boxed system, but thought that this system could be made more transparent by an HP wireless device that would allow students to see into classrooms, libraries, and research centers and to surveil the locations of students and professors alike. In contrast, one might understand the first-year college experience nostalgically as a once-in-a-lifetime liminal time between childhood and career that must be captured and archived for posterity with a very different device, the Apple iPod. Finally, the recent romance with paperless education and QR labels represented by iPad distributions merits further close reading. As we unveil new interfaces for the classroom, memories of other experiments with the mobile screen should not be erased, and the longer history of instructional technology, participatory culture, and embodied interactions—rather than the latest rhetoric from a product showroom—should provide our reference points.

8 The Play's the Thing: Games and Virtual Worlds in Higher Education

At the University of California, Irvine (UCI) in 2007, I taught an upper-division course in social media that held some of its pedagogical activities on Anteater Island in Linden Labs' 3D computer-generated online world Second Life (SL), where participants are represented by avatars and transportation is commonly effected by flying or teleporting. Thanks to support from university administrators, indoor and outdoor classrooms had already been constructed on the island, which was named after the university's mascot, along with an elaborate conference center with space-age decor. Plans were underway to add more audiovisual equipment to the island's basic facilities. For the first time in my academic career, I had to consider the question of whether I wanted to make a claim to property ownership as part of merely planning a syllabus for a class. Sometimes coordinating the most basic support services felt surreal. I remember consulting a magical fox-like creature busy with the university's expansion plans just to check classroom scheduling.

Naturally, given the nature of the university's purchase, my course was not the only one on campus that was taking advantage of UCI's recent acquisition of this virtual property. Classes in game development devoted space to building 3D prototypes. Computer science students were tinkering with the SL scripting tools, which instructors hoped would teach basic principles of coding. In restricted space on the island, cordoned off by no-admittance banners at the shoreline encampments for these other classes, there were pup tents marked with American flags and skyscrapers reminiscent of some of the surrounding real-word research parks in Orange County. Elsewhere on the island, enthusiastic librarians were trying to figure out how to incorporate plans for Anteater Island into their other online services that connected users seeking research advice with physically remote librarians. They described themselves as inspired by the work of James Paul Gee and

Henry Jenkins and championed learning in game worlds and the fostering of digital literacies.

Even before the acquisition of Anteater Island, several of my colleagues had been using Second Life as a site for data collection and experimental testing. For example, professor of anthropology Tom Boellstorff had explored its social landscape for his book *Coming of Age in Second Life: An Anthropologist Explores the Virtually Human*, which documents the experiences of citizens living, working, and playing in alternative online realities.[1] Boellstorff was interested in the site as a place for training anthropology students and for offering language-learning practice with native speakers. Professor Crista Lopes, from the Department of Information and Computer Science, had invented a wearable browser with possible commercial applications for online retailers. She also promoted the idea of a "universal campus" and was testing different configurations for teaching and research activities.[2]

My objective in using Second Life was different from others at my university. In some ways it was to be a proving ground for the readiness of Anteater Island for general education courses. In contrast to disciplinary efforts focused on testing research questions, I planned to use Second Life as a rhetorical space for teaching fundamental communication skills, public speaking, and visual presentation techniques. I wanted students to take advantage of the ways that barren environment functioned as a kind of state of nature—one in which little could be done without self-consciousness of material scarcity and the need to improvise with available digital tools. Guest experts who visited my class took the students out of Anteater Island to explore more elaborate properties used by philanthropic organizations, health agencies, political activists, and government institutions and explained possible public-relations rationales for using virtual worlds to reach real-life audiences.

Of course, I wasn't the only writing instructor holding classes in Second Life. Sarah Robbins at Ball State taught composition in a Second Life virtual environment. One of her students reporting on obesity had adopted the avatar of the Kool-Aid Man for an in-class presentation, and another had created a giant virtual sculpture to protest high tuition costs.[3] Lisa Gerrard at UCLA used Second Life in her first-year writing courses; she has a background teaching writing in older text-based online environments, such as MUDS (multi-user dungeons, sometimes also known as multi-user dimensions or domains) and MOOs, which are object-oriented MUDS.[4] Although rhetoric has had a long history of nonverbal communication, I was leery of making 3D modeling and animation required assignments, so our sessions in Second Life were limited.

In the 2006–2007 academic year, UCI was not alone in embracing this new forum for potential distance-learning projects. After Second Life was augmented with sound for live audio chat, another significant barrier to using it as a learning space had been lifted. Although much of the electronic discourse generated by institutions of higher education took the form of 2D static posters and PowerPoint electronic slideshows that did not really use the 3D spatial presentation of the site, high hopes existed that virtual worlds would be able to reach broad constituencies. Campuses planned to serve students who couldn't be physically present on a university campus, who had anxieties about learning alongside much younger—or older—students in their physical-world identities, or who were just excited about being early adopters of new technologies. Colleges found a range of uses for Second Life—from hospitality-industry students in Hong Kong who earned virtual goods as they completed learning benchmarks to literature students in Pennsylvania who met in the landscapes of the novels they were studying. Psychology faculty hoped to create virtual laboratories where interaction could be studied more objectively.[5] Even the most august institutions were participating in Second Life land speculation. By using the transmedia group dynamics of the online video-sharing site YouTube, Harvard University Law School professor Charles R. Nesson recruited students from the general public to fill his cyber-law course. Enrollees through the university's extension program attended Nesson's classes on Berkman Island, on terrain named for the university's Berkman Center for Internet & Society. Imaginative designers were free to create, in the absence of weather, buildings without roofs—as well as buildings without stairs to accommodate users who could fly. Nonetheless, Second Life campus architects often recreated historic landmarks and existing structures that were exact copies of those already on their own campuses.

Some faculty saw Second Life as an alternative to more costly "cave" installations or applications that required users to wear awkward and fragile head-mounted 3D displays. Chemistry professor Jean-Claude Bradley created 3D molecular models for his Drexel University organic chemistry students. Although only about 5 percent of his students visited the optional exhibit, Bradley felt his efforts on his campus's dragon-shaped island were worth his academic labor. He told a reporter from *USA Today* that Second Life allowed his students to interact with faculty and each other differently. "I can show them molecules in three dimensions. We can walk around the molecule and discuss it."[6] Similarly, Jacquelyn Ford Morie of the Institute for Creative Technologies at the University Southern California (USC), which has a national reputation for creating 3D environments, such as

Virtual Iraq, for training military personnel by integrating head-mounted displays and licensed 3D game engines in their software projects, used Second Life tools to create far less expensive models of Iraqi villages in which soldiers could practice scenarios and learn the rules of engagement.

Certainly, there were ambitious construction projects underway while I was teaching in Second Life. Environments were lovingly crafted, with hundreds of man-hours, to promote learning and help students retain course material. For example, Steven J. Taylor, director of academic computing at Vassar College, recreated the interior of the Sistine Chapel on the college's island, where visitors could step inside a lemon-colored edifice and view a miniature simulacrum of the frescoes that adorn the fifteenth-century chapel in Vatican City.[7] Since avatars could fly up to the ceiling to get a better look, in some ways it was a superior learning environment to the original. However, if leasing land from certain developers with particular zoning restrictions, universities could find themselves limited architecturally. The guide for the Annenberg Island tour of USC's virtual center for public diplomacy in Second Life explained that Asian-themed architecture was mandated for any approved construction.[8]

In Second Life, instructors had to deal with unfamiliar problems involving codes of hospitality, which ranged from the profound philosophical questions rehearsed by Lévinas and Derrida to the most mundane rules of admittance. After all, anytime a stranger comes into university space, it can be complicated to manage being welcoming. In a traditional classroom, even the arrival of a sibling or friend can activate faculty territorial feelings. In addition, sometimes the mentally ill or the homeless will perceive campus buildings as part of public land. Unlike trading digital materials designed for a particular classroom or lecture hall that may be shared as PowerPoint files or HTML documents with colleagues, sharing space in Second Life could actually interrupt precious class time. Nonetheless, educational artifacts were often explicitly designed for public consumption and shared use. Whether it was a giant revolving molecule or a Sistine Chapel full of virtual frescoes, educators using Second Life extended hospitality to colleagues who might express interest in using their virtual buildings or classroom aids with their own classes of students, and social capital was aggregated through sharing.

Sometimes Second Life instruction seemed to destabilize disciplinary divisions, and online mingling appeared to foster opportunities for more interdisciplinary experimentation. For example, as her avatar Max Chatnoir, biology professor Mary Anne Clark mixed instruction in genetics with music on Genome Island, where students from Texas Wesleyan University experienced the way in which she made digital music by correlating amino acids in a protein—such as those derived from spider-web silk—to musical notes.

It should not be surprising, however, that not everyone was enthusiastic about the use of this technology for distance-learning purposes. In *Digital Diploma Mills*, David F. Noble castigated online education as the reincarnation of exploitative and ineffective correspondence schools in the nineteenth and twentieth centuries that promised access for nontraditional learners but often prioritized profit over pedagogy.[9] In revisiting Noble's criticisms, digital theorist Mark Nunes expresses more hope for the promise of the virtual classroom, based on his previous pedagogical experiences in MOO environments. However, in his book *Cyberspaces of Everyday Life*, Nunes argues that online environments resist simplistic moralizing, because they manifest multiple ideologies: "The rhetoric of access and control converge in the production of a networked social space that 'center[s] learning around the student instead of the classroom,' or, rather, that disarticulates the classroom as a space of learning, locating it in the mobile, privatized, and irruptive spaces of control articulated in the lived practice of the 'student body.'"[10]

Even among digital researchers who had been at the forefront of using virtual environments for teaching, there were many skeptics about Second Life. At a 2007 panel at the annual SIGGRAPH convention for computer graphics professionals, Georgia Tech professor Amy Bruckman argued in the presence of Linden Labs executives that 3D worlds were poorly designed for pedagogical use. Although many were ecstatic about the instructional possibilities of "end-user programmable worlds," Bruckman charged that Second Life was inappropriate for many kinds of pedagogical interactions and that using technologies with which students were already familiar—such as social network sites and text messaging on cellular telephones—made more instructional sense, if the aim were to provide more student-centered teaching.[11]

As early as 1995, Bruckman had cautioned against thinking about cyberspace from the viewpoint of a virtual tourist interacting with a themed environment designed to be appropriate for all users.

> Cyberspace is not Disneyland. It's not a polished, perfect place built by professional designers for the public to obediently wait on line to passively experience. It's more like a finger-painting party. Everyone is making things, there's paint everywhere, and most work only a parent would love. Here and there, works emerge that most people would agree are achievements of note. The rich variety of work reflects the diversity of participants. And everyone would agree, the creative process and the ability for self expression matter more than the product.[12]

She observed that cyberspace was constituted by the ephemeral practices of users rather than rigid walls and ceilings in the logic of a Cartesian grid. Thus, she envisioned a messy environment of trial-and-error juvenilia very different from the elaborate themed exhibits of miniature Sistine Chapels or spinning molecules featured in university press releases.

Bruckman's point—that cyberspace is not Disneyland—would prove to be important to administrators concerned about the large user population in Second Life of erotic explorers interested in enacting sexual fantasies. Often this experimentation was comprised of simple cross-dressing, but role-playing could also include taboo activities around S&M, slavery, or child molestation. Although these areas of Second Life were impractical for teaching, university Internet researchers jumped at the opportunity to observe shadow practices. For example, Shaowen Bardzell studied the lives of sexual submissives,[13] and Kinsey Institute researchers looked at the site to gather data about a variety of aberrant practices involving nonhuman participants.[14] Others did fieldwork in popular parts of Second Life that were filled with Goreans—followers of John Norman's science-fiction fantasy novels—who participated in elaborate, ritualized master-slave groupings. Even before the existence of 3D virtual worlds, Julian Dibbell had documented rape in the text-world cyberspace of LambdaMOO.[15] Soon the problem of "avatar rape" in Second Life instructional spaces became featured in higher education press coverage,[16] and many wondered how to apply their campus's sexual harassment policies in a corporate-owned hedonistic virtual world.

Although the metaphors of crowds and crowding (smart mobs, swarms, hives, etc.) were often used to describe online participation during that period, large parts of Second Life were empty, which made it appealing as a site for sexual intimacy and subcultural practices. Conversely, this same privacy might also be conducive to some pedagogical situations. Students, who had become increasingly likely to seek out their professors online through e-mail or chat, might be more comfortable consulting faculty in virtual office hours. One of the classrooms at Anteater Island, in fact, projected this binary into its design scheme: it was a "holodeck" built by "Anteater Beaumont" that could be transformed into a library, classroom, and seminar space. But strangely, given the fact that it was ostensibly university property, it also could shape-shift into configurations for "bedroom," "dinner for two," and a bar and disco as "club 360" with a VIP Room.

Long before Bruckman's cautionary SIGGRAPH message, others who studied the history of digital culture were expressing reservations about how the graphics-intensive computing demands of virtual world technology reinstituted a digital divide that had been disappearing elsewhere, as flatter, text-based forms of cyberspace had proved more appealing. The elite and idiosyncratic user base also differed dramatically from the demographics of the general population. Clay Shirky spurred a series of lively exchanges in the academic blogosphere when he indirectly initiated a debate with Henry Jenkins by comparing Second Life to the by then little-used LambdaMOO. Shirky wrote, "If, in 1993, you'd studied mailing lists, or Usenet, or IRC, you'd have a better grasp of online community today than if you'd spent a lot of time in LambdaMOO or Cyberion City."[17] To Shirky's criticism of Second Life, Jenkins responded that during the "Renaissance and the Age of Reason… key innovations occurred among a much smaller number of artists and thinkers" so that a "small community of people can generate an enormously rich culture and can have a transforming impact on society as a whole."[18] Nonetheless, even Jenkins acknowledged limitations: "The numbers matter if we are asking whether Second Life represents 'the future of the web' but personally, I have never believed that SL is going to be a mass movement in any meaningful sense of the term. … I do not buy the whole nonsense that immersive worlds represent web 3.0 and will in any way displace the existing information structures that exist in the web, any more than I think audio-visual communications is going to replace written communications anytime soon."[19]

Furthermore, the manipulation, or "puppeting," of avatars was very attention-intensive and required considerable practice to do adroitly. Of course, students might have learned valuable lessons from these situations of discomfort and constraint, much as they have learned from successful exercises simulating blindness in the history of pedagogy, but there might be better ways to question ideologies of organic naturalism in user design. Students were constantly reminded of their inabilities to perform the simplest operations associated with verbal and nonverbal communication. As an instructor, I found it could be difficult to have successful class participation, since students struggled to interact with faculty, guest speakers, and each other on the site. It often felt like teaching a class of preschoolers who were likely to wander off or succumb to strange bouts of fidgeting. The online course management company Blackboard seized on these obvious limitations and produced a parodic video mocking Second Life to promote their own online instructional services.[20] Footage of live-action Blackboard workers showed them acting out the roles of avatars who awkwardly struggled to get dressed, open classroom doors, sit down at a table in a conference room, or derive answers to basic questions.[21]

Universities also found themselves policing forms of vandalism, for which they were extremely ill-equipped, and grappling with much more complicated questions about academic freedom than those associated with protests on in a traditional college space. For example, in 2010, Woodbury University pulled its virtual campus out of Second Life after being banned by the company for the second time for failing to control griefers associated with the Patriotic Nigras, an SL group known for racist and homophobic slurs, and a contingent from the prankster forum 4chan.[22] Northwestern University philosophy professor Peter Ludlow covered the back-and-forth hostilities and often seemed to express sympathy for the faculty advisor and his students. The Woodbury contingent was attempting to infiltrate the Justice League, a kind of virtual neighborhood-watch group, which had reported the university for tolerating online outlaws. Academic freedom issues for subscribers of Second Life services in higher education could be tricky, since institutional users were subject to what could seem to be arbitrary exercises of power. Linden Labs could raze entire building complexes and bar teachers from their classes without notice; they claimed that such actions were justified by long click-through contractual agreements common among Internet companies.

On March 1, 2011, UCI's Anteater Island was closed down. A press release from the library explained that the decision "to cancel this virtual world software platform was made after Linden Labs, its provider, will no longer honor educational pricing and the renewal price doubled."[23] The library also reported "a decrease in new participation by faculty and students." It is easy for critics of Second Life to argue that the higher-education pullout was inevitable, but it may be more useful to think about what this episode in instructional technology tells us about ongoing investments in online learning and teaching that might be less obviously flawed.

Fashioning Knowledge

The use, study, and meaning of fashion has often been underrated in the academy, although in recent years there has been more explicit discussion about the importance of apparel in postsecondary contexts. Poststructural critics who followed Roland Barthes were some of the first to attend to costuming. In a 2006 Modern Language Association talk, Elizabeth Deeds Ermarth referenced the cultural conversation taking place about fashion in the academy and argued that it provided yet another system of codes to be recognized by deconstructionist differentiation.[24] Fashion had also become important to literacy theorists like James Paul Gee, who characterized literacy as being "an identity kit" constituted like a dress-up kit made up of separable pieces that mark, disguise, and costume the wearer.[25] Blogs and other new forms of publication instigated more discourses about the importance of clothing in higher education. Of course, feminists had been concerned with how fashion could function as both a verb and a noun for decades. Thus, in academic contexts, fashion could both discipline subjects and establish shared communities around gender identity. For example, in her work on "Silhouettes for the 21st Century," media artist Heidi Kayser noted that fashion can be "a prevailing style in custom, dress, or behavior…

Figure 8.1
Poster about a workshop with Sarah Robbins for teaching in Second Life. Courtesy of Sarah Robbins.

to train or influence yourself into a particular state or character," or "to adapt, as to a specific purpose."[26] In virtual worlds, as in the embodied sociality of everyday life, fashion shapes how one's character is performed.[27]

In Second Life, fashion was everything. Outfitting one's avatar was generally the primary thing to be done as a first-time Second Life visitor. The Lindens eagerly reported on entire online economies devoted to rewarding the elaborate scripting practices that could build embroidered purses, hair accessories, or other examples of clothing as wearable art. As Michele White has pointed out, avatar outfitting had a prehistory in the 2D world of social media and online chat that even generated its own anxieties about plagiarism, appropriation, and the tentative nature of the ownership of intellectual property online.[28] For instructors teaching in Second Life, one workshop about pedagogy in virtual worlds emphasized that it was critical not to come to class naked, either literally or, if virtual clothing was poorly chosen or rhetorically ineffective, figuratively.

It was true that in Second Life some academics, such as Henry Jenkins and Lawrence Lessig, appeared much as they did in their face-to-face teaching and public advocacy roles. In Lessig's case, his avatar appeared in a black pullover and black collarless jacket, just as he was attired when evangelizing in person for free culture; Jenkins' avatar wore jeans, a striped shirt, and his distinctive suspenders when engaged in outreach activities for his participatory culture partnerships. However, because these avatars often represented younger and thinner versions of their real-world professorial selves, these self-representations still marked a specific mode of dress for online interactions as they adopted the informal persona of "professor avatar."[29]

A significant contingent of Second Life faculty went much further in cultivating glamour in their online attire. They embraced the opportunity to make sequins, robes, and even wings a part of their academic wardrobes. Trading such objects of apparel in online worlds, including what Celia Pearce described as "twinking" or gifting, served many important social purposes, particularly for establishing friendships and initiating new members into groups.[30] Pearce, as a representative of the Ludica feminist game collective, claimed that, even in combat-oriented role-playing games, experimentation in constructing gendered identities through recombination of attire had a variety of important functions. Academics with impressive regalia clearly were experienced users who could either work effectively with Second Life's 3D design interface or had the social capital to have others gift numerous items of clothing to them.

The medieval period has a special significance for inhabitants of many virtual worlds and games, and it has been important in shaping many aspects of the traditional university as well. Academic rituals, such as PhD hoodings and college graduations, remind participants of the importance of academic finery as a signifier of their institutional membership and rank, and many of these particular costumed dramas go back to the founding of the first universities in Europe. During the medieval period, luxury goods were an important indicator of class, access to exotic foreign markets, and one's ability to marshal labor within a given domestic or craft culture, so it was not surprising that, as new institutions of knowledge were being founded, they embraced clothing as a marker of status and created spaces for the pageantry of costumes of expertise and allegory.

However, Henry Jenkins claimed that the hierarchy of Second Life, in which clothing was a marker of specialized knowledge or elite membership, was easily capable of being subverted: "Some have dismissed SL as a costume party—I see it more as carnival in the medieval sense of the term—as a time and place within which normal rules of interactions are suspended, roles can be swapped or transformed, hierarchies can be reordered, and we can step out of normal reality into a 'magic circle' or 'green world' which can be highly generative for the imagination." [31]

For instructors who are in charge of credit-bearing courses in Second Life, the thought of holding a costume party with a distance-learning dimension could seem a daunting pedagogical challenge. And yet costume dramas still play out all over the university grounds, from sober graduation ceremonies to political buttons and t-shirts worn at noisy demonstrations. Even though instructors may have disciplined themselves not to see their students' and colleagues' fashion choices or acknowledge their identifying codes of dress, partly out of justifiable respect for their university's stated harassment policies, Second Life made this intentional blindness nearly impossible. Despite pressures to maintain the humbleness and uniformity of dress that signifies ignorance of commerce and material concerns, fashion performs a number of important tasks in communicating information and expressing identity.

Unfortunately, there was also a less flattering way that Second Life emulated the feudal past, in that—for all the surrounding talk of virtual capitalism and futuristic entrepreneurialism associated with this venture by Linden Labs—users agreed to property restrictions and labor conditions that relegated them to the economic and cultural role of vassals, if not serfs.

Badges for Learning

Although some instructors still teach in the virtual worlds created for online games, such as *World of Warcraft* or *Minecraft*, today's status markers in alternative forms of online education tend to emphasize 2D badges rather than 3D costumes and objects. Badges and other icons of achievement became important in games played on social network sites, such as Facebook, but they had an even longer history in digital culture as signifiers of in-world experience, skill, social capital, reputation, and negotiating ability, as well as in predigital cultures of scouting and military service. As designed virtual objects, digital badges often use pixels extremely efficiently to maximize legibility and to communicate a strong visual aesthetic. Such badges can serve as collectible assets in a user's inventory, and—unlike points that serve as a single unit of currency—they often explicitly foster variety in user behavior.

The badge concept gained more currency among educators after the 2007 Presidential Address to the American Educational Researchers Association recommended that high-stakes points-based testing in a few subject areas, which was established at the federal level by No Child Left Behind legislation, be replaced with a new system oriented toward "acquiring concrete Qualifications that certify important accomplishments" to ensure "a wide choice of Qualification areas."[32] Advocates for reforming higher education also became interested in the badge concept; efforts included badges given out in places such as the online university P2PU for their School of Webcraft courses,[33] the University of Michigan's Open Educational Resources program (to encourage sharing of pedagogical materials among educators),[34] and the Khan Academy.

Software companies and civic organizations, including 4H and the Girl Scouts of America, supported the concept of the Mozilla Open Badges project, which was devoted to developing the open specification and application programming interfaces that would allow any participating organization to offer badges in a standard, interoperable manner.

> A digital badge is an online representation of a skill you've earned. Open Badges take that concept one step further, and allows you to verify your skills, interests and achievements through credible organizations. And because the system is based on an open standard, you can combine multiple badges from different issuers to tell the complete story of your achievements—both online and off. Display your badges wherever you want them on the web, and share them for employment, education or lifelong learning[35]

As in the case of conventional scout badges, "earning" rather than "learning" becomes the key verb in the process. The rhetoric of transparency and openness is also pervasive, although—as I have argued elsewhere—openness can be difficult to define. Even the "open standard" designated as key to technical specifications can be up for debate.

Attention to badges within the higher education community intensified when the Humanities, Arts, Science, and Technology Alliance and Collaboratory (HASTAC) announced in September 2011 that it would administer a "Badges for Lifelong Learning" competition with two million dollars worth of award money and sponsorship from the MacArthur Foundation, the Mozilla Foundation, and the Gates Foundation. Although they were working with Mozilla, the HASTAC definition emphasized that badges would serve instrumentally as tools that can indicate evaluative measures rather than merely as representations that display blanket certification.

A digital badge is a validated indicator of accomplishment, skill, quality, or interest that can be earned in many learning environments. Today more than ever, traditional modes of assessment fail to capture the learning that happens everywhere and at every age. Digital badges are a powerful new tool for identifying and validating the rich array of skills, knowledge, accomplishments, and competencies.[36]

Like the "credible organizations" required by Mozilla, the HASTAC call specifies that there must be a mechanism for a "validated" evaluation. Although evidence of mere interest seems to mandate a relatively low threshold of accomplishment in comparison to skills, both Mozilla and HASTAC adopted this inclusive approach. Given the open nature of the call, almost a hundred finalists advanced to the final round of judging.

More than thirty winning projects were announced in March 2012. The winners included a number of government agencies, corporations, and academic institutions. The University of California at Davis received a $175,000 grant for adopting badging in conjunction with an electronic portfolio system for undergraduates in their Sustainable Agriculture and Food Systems major.[37] Kevin Carey, policy director at the think tank Education Sector, praised the large land-grant campus for jettisoning traditional distribution requirements, seeking a remedy for grade inflation, separating credentialing from teaching, and being open to input from parties outside the university who might take over credentialing duties for the badge one day, such as "farmers, students, civic groups, companies, professional organizations, and individual scholars."[38] The University of Southern California planned to use their grant money to enhance their service-learning programs with badges, since "the service-learning community has struggled with how to identify and recognize the outcomes that students learn, like civic knowledge and diversity."[39]

As badge programs spread to certificate-awarding organizations like OpenStudy and MITx and galvanized more people to condemn "the tyranny of a degree," the potential conflict with diploma-granting institutions and their faculty was exacerbated. Critics objected that badge systems might only magnify the worst qualities of existing test/grade/diploma systems that reify and commodify educational objectives rather than nourishing learners' development. In "Welcome to Badge World," Alex Reid warned that the emphasis on getting to "level up" and "make things count" would only extend the "god-awful experiences" created by existing high-stakes testing "to every waking moment of your lives."[40] Much as so-called casual gaming can be, because of the simplicity and omnipresence of its appeals,[41] remarkably engrossing, casual badging could make snackable educational experiences for meaningless virtual objects the new obsessive norm for learning. Reid also opined that such initiatives, in the absence of centralized oversight from professional and institutional stakeholders accountable to accrediting bodies, would empower digital diploma mills to crank out meaningless accolades for those who paid generously for their services.

Researchers also cautioned that badges should not be thought of as "rewards." In real-world reputation economies with robust badging for skills such as programming in particular computer languages, the actual functions of badges had proved to be much more complex. Earning badges in these contexts required that standards be established through patterns of use and common conventions, such as adopting recognized genres of testimony from experts attesting to competence or agreeing on familiar routes of making progress, so that those who would interpret the meaning of a given badge were aware of its conditionalities and entitlements.[42]

Those pointing to the limitations of badge systems were also building on decades of studies done by educational psychologists showing problems with relying on extrinsic rewards rather than intrinsic motivations. Experiments testing the "overjustification effect" examined the counterintuitive impacts on learners' motivation caused by awarding recognition with gold stars, candy, money, and other tokens of achievement. In one of the most famous studies conducted in 1973, preschool children were shown to be considerably less likely to freely choose drawing as an activity if they had received a ribbon for doing so earlier.[43] Encouraging incentives actually discouraged the rewarded behavior. A 1971 famed study by psychologist Edward Deci showed that college students who were given a dollar for completing a cube puzzle were less likely to freely devote time to that activity than were students in an unpaid control group.[44] He subsequently published a significant body of work with Richard Ryan devoted to self-determination theory that demonstrated that students tasked with many different activities, from problem solving to written expression, could be inhibited from pursuing educational goals associated with rewards.

Resisting Gamification

In a 2011 speech called "We Don't Need No Stinkin' Badges" at the Serious Games Summit, game creator and motivational speaker Jane McGonigal castigated proponents of gamification for ignoring participants' desire for *intrinsic* forms of satisfaction rather than *extrinsic* rewards.[45] She was referring to what had become one of the big buzzwords in corporate sales, digital entertainment, and even classroom teaching: gamification—a term that describes the application of game-design principles to nongame contexts.

In recent years, gamification has become inexplicably trendy in higher education. Novice faculty receive e-mail invitations to seminars about how to "gamify" or "gamifi" their courses.[46] The work of professors tackling the "gamification of college lectures" is covered in laudatory press releases.[47] Workshops at professional conferences devoted to teaching and learning provide tips on revising whole curricula to meet gamification objectives.[48] Course catalogs advertise classes in gamification, and colleges even offer certification in "gamification and serious games."[49]

A tech start-up entrepreneur claims to have coined the term "gamification" in 2002,[50] but it seems to have appeared independently in several fields, including management studies and game studies, before becoming, around 2010, a topic of discussion in the research and development of "serious games,"[51] which are games primarily designed for purposes other than entertainment that aim to accomplish goals such as training military personnel, preparing for disasters, reinforcing healthy lifestyles, raising political awareness, or achieving learning objectives in education. A major manual for the gamification movement was the 2009 how-to guide for corporate executives *Total Engagement: Using Games and Virtual Worlds to Change the Way People Work and Businesses Compete,* which was coauthored by a venture capitalist/CEO and a Stanford games researcher.[52]

Ironically, McGonigal had appeared as a keynote speaker at the Gamification Summit a few weeks earlier, during a launch-day event for her bestselling book *Reality Is Broken,* which has since become another bible for the gamification movement. In her more critical Serious Games Summit talk, McGonigal disparaged gamification as superficial and differentiated it from what she calls "gameful" approaches to activities that harvest collective intelligence. In her positive-psychology approach to applying the design of play experiences to real-world problems, happiness serves as an important end in itself, and games are a way to get there. As proof of this, she praises the massive coordinated participation of volunteer labor in online game worlds, such as *World of Warcraft,* and lauds the massively multiplayer protein-synthesis game *Foldit*, which predicts that medical breakthroughs will result from players' labor, and the game *Free Rice*, which promises to "donate 10 grains of rice" for "each answer you get right."[53] According to McGonigal, these gameful activities have an altruistic dimension that is different from more ego-oriented gamified reward systems. Her vision of a happier, healthier, more positive, and more prosperous world promoted through games also depends on free consent, unlike the real-world incentives luring participants to gamification, which may coerce users to give their labor under duress.

In the more caustically worded "Gamification Is Bullshit," fellow designer and critic Ian Bogost has argued that most of the basic premises of the current gamification movement differ little from those of conventional brand-marketing to consumers to reinforce buying habits and have little to do with the sophisticated feedback systems of compelling games. Business gamification advocates also read game culture too superficially by "mistaking incidental properties like points and levels for primary features like interactions with behavioral complexity" to maintain "reassuring" and "easy" practices of salesmanship rather than acknowledging that "games can offer something different and greater than an affirmation of existing corporate practice."[54] When the player's labor is completely co-opted by gamification, Bogost characterizes it as "exploitationware."

In tackling the challenges of reforming higher education, both McGonigal and Bogost agree that games could provide useful models for learning, critical thinking, and engaging in collaborative activities. Although they feel that gamification is a bad model, they believe that rule-based systems have a strong potential to teach students much more actively by having them engaged in hands-on doing that allows ideas to be tested dynamically. McGonigal's catalog of the components of a game (goal, rules, feedback system, and voluntary participation) maps well onto the elements of her ideal educational approach. Bogost claims that developing "procedural literacy" provides a paradigm for other forms of critical reasoning. For him, players figure out how to play and win games without manuals that enumerate all the rules explicitly, just as novices in academic subjects advance to expertise by deducing principles from observed phenomena, and instructors figure out how to succeed at their professions by interpreting feedback from their actions.[55] Games involve values and aspirations that are major components of meaningful education as well, according to Bogost.

In *Persuasive Games*, Bogost acknowledges the influence of literacy theorist James Paul Gee, who is famed for arguing that video games constitute a model for effective learning, because they allow risk taking, learning from failure, exploration of identity, and other give-and-take experiences that characterize successful classrooms.[56] Ironically, much of Gee's much-cited pioneering work argues for the educational value of *commercial* games, rather than those designed specifically for classrooms. He asserts that video games, such as *Tomb Raider* or *Deus Ex*, provide opportunities for "situated learning" or "embodied learning" and a pathway to higher-order forms of literacy. Yet Gee warns that games can be difficult for conventional educational institutions to incorporate, because learning and social transgression are closely related in video games.[57] Furthermore, Gee claims this literacy operates through the acquisition of general principles for social interaction and problem-solving—not through memorizing discrete sets of facts or repeating specific skills. Bogost argues that in many ways Gee's position is not strong enough, in that video games "offer meanings and experiences of *particular* worlds and *particular* relationships" rather than just general operating principles. As Bogost explains his thesis about educational video games, "The abstract processes that underlie a game may confer general lessons about strategy, mastery, and interconnectedness, but they also remain coupled to a specific topic."[58]

Should Games Make People Happy?

McGonigal and Bogost fundamentally disagree about how games should make people feel. While McGonigal insists that games promote happiness, Bogost argues that games should promote critical awareness instead. Bogost is known for creating and admiring games that situate players to face what he calls the "rhetoric of failure" by creating a game mechanic that "comments upon a political situation by denying players a victory condition."[59] In contrast, McGonigal promotes a gospel of success based on "happiness activities" that serve as "the daily multivitamins of positive psychology," which have been "clinically tested and proven to boost our well-being" and "backed by multiple million-dollar-plus research studies, which have conclusively demonstrated that virtually anybody who adopts one as a regular habit will get happier."[60] As simulations that recreate the routinized dynamics of the classroom or lecture podium, Bogost is interested in games that provide cynical portrayals of dysfunctional educational processes, such as the 1975 game *Tenure* or his own 2008 game *Honorarium*. In contrast, McGonigal turns an academic research area into a triumphant superhero in her cheerful 2004 game-based project *PlaceStorming*.

Sometimes McGonigal and Bogost will read the same game quite differently. For example, Bogost says that his games about travel regulations are meant to educate the player about the existential futility of tourism and airport security rules and "amplify their estrangement," but McGonigal claims that such games remind players "of our power to improve our experience"[61] and provide a way to make travel "more fun and personally satisfying."[62] McGonigal characterizes an assassination game about public politeness that is played with mobile phones, *Cruel 2B Kind*, as a vehicle for "the basic happiness activities of expressing gratitude and practicing random acts of kindness";[63] Bogost emphasizes the ironic contradictions of killing with kindness and the moral ambiguities of assaulting innocent bystanders.

In general, Bogost becomes skeptical when experimental game designers want to "explore or arouse positive emotions in their players" and argues that designers "sell themselves short with this trite incantation about emotions."[64] Often, when Bogost is discussing that which is "happy" in a game, he means the term in a very particular technical sense, one derived from the speech-act philosopher J. L. Austin, who defines "happy" and "unhappy" performative actions in the context of a particular happenstance. For example, Bogost sees a game experience as "unhappy" if "a game performs an action without the player's understanding of its implications," because "it confuses performativity and exploitation."[65]

After the appearance of *Reality Is Broken*, Bogost publicly criticized McGonigal's assertions that a "broken" reality can be "fixed" by the collective efforts of gamers and that games were "happiness engines." Although he praised her collegiality and friendship, he questioned her vision of the ultimate purpose of games and gameful living.

> For me, the solutions we find through games do not lead us to more successful mastery of the world, but a more tranquil sense of the elusiveness of that mastery. The systems-thinking games embrace shatters the very ideas of world-changing with which we have become so accustomed. And we don't occupy game worlds because the real world isn't happy or fun enough, but because we need help embracing that real world through the properties of ambiguity and intricacy that make games like the world in the first place.[66]

In a similar vein, Jesper Juul insists in his recent book *The Art of Failure* that to think of games as "fun" is to misunderstand them profoundly; Juul observes that comparisons to tragedy might be more appropriate as a way to understand the cultural function of games.[67]

Although the vacuous phrase "serious games" has become another buzzword in higher education, good games can inspire real gravitas, as the best independent games do, particularly when they raise painful issues about unworkable assumptions or contradictory rules. I agree with Bogost that academics sell themselves short when they buy into an ideology of edutainment, and they sell their students short as well. Treating education as a serious matter—even if satire and subversion are part of the experience—shows that we take our students seriously. Edutainment infantilizes undergraduates; it underestimates their willingness to commit to difficult intellectual work and trivializes the ambitions of their families by making it sound as if young people have only come to campus to have fun. Making students wise is much more important than making them happy. Happy people can ignore global warming and the risk of nuclear annihilation. We need to develop citizens capable of understanding hard choices and complex problems, which the best kinds of games—like the best kinds of novels or films—can do.

Should Games Be Fun?

On October 2, 2007, the *Chronicle of Higher Education* reported that the MacArthur Foundation was pulling out of an ambitious plan for a multiplayer game that would teach digitally savvy students by presenting Shakespeare's works in a 3D virtual world full of opportunities for interactivity in the form of questing, crafting, card playing, and answering questions in trivia games.[68] As soon as this funding cut-off to the Synthetic Worlds Initiative at Indiana University was announced, it stimulated hand-wringing throughout the serious games field about the viability of the entire educational video game movement.

After the MacArthur Foundation refused to renew funding for *Arden*, the director of the project issued a public *mea culpa* about the project's failure on a prominent, collaboratively written blog about virtual worlds. As a political economist who had studied massively multiplayer online role-playing games (MMORPGs), he had hoped to use *Arden* as a social laboratory to test different economic models and find insights into the way that money works in the real world. Unfortunately, as Edward Castronova explained in diagnosing the failure, his game's function as an economic simulation was insufficient to justify its existence. He summed up his perspective on the problem simply: "It's no fun. ... You need puzzles and monsters."[69]

Arden raised many organizational and managerial issues, such as those concerning the limits of the expertise of the university, play's relationship to philanthropy, and the circulation of units of value in real and virtual worlds.[70] This "failure" also bears on philosophical questions about game design, particularly those related to the principles governing the adaptation of literary works. Furthermore, for "narratologists" working on *Arden*, video games were seen primarily as representations of dramatic action that featured characters, settings, and plots. For "ludologists," video games were experiences that involved interacting with rules. Those who focused on questions of narrative viewed the game's pedagogical mission one way, while those who focused on procedurality viewed it in another. Focusing on aesthetics or ethics might also influence how participants saw their advocacy roles in the game-development process.

For decades, promoters of instructional technology have argued that such games could provide more learner motivation and better measurement of pedagogical objectives by incorporating play more easily to fulfill the obligations of formal education. Thus, video games had been designed for military drills, preparation of emergency first responders, disease prevention, patient rehabilitation, sensitivity training, conflict resolution, and teaching and learning at all levels, even though—like other gamers—players often "cheated" once the affordances of a system had been learned.[71]

However, even during the heyday of exuberant optimism in the educational games movement, many affiliated with game studies were already providing more nuanced readings of the potential value of games to learning. They cautioned against uncritical acceptance of the sales pitches of one-sided boosters who thought that any learning objective could be combined with any video game, which Scot Osterweil characterized as the "Grand Theft Calculus" mentality.[72] Kurt Squire and Henry Jenkins also claimed that game skills did not necessarily transfer fluidly to real-world environments, although games helped users participate in the social practices that were constituents of genuine learning; they also pointed out that recreating literary works in video games should not collapse the differences between games and traditional storytelling media.[73]

For perspective, it is worth remembering that *Arden* was not the only educational video game ever created to teach Shakespeare, so its "failure" might not be the last word on the subject. In Canada, English professor Dan Fischlin created a scrolling shooter game, *'Speare*, that encouraged players to examine individual text fragments from the Bard to better understand the play *Romeo and Juliet*. The plot centered on the elaborate backstory of a "society based on knowledge and poetry," with two warring planets—Capulon and Montagor—in the Verona system working together in order to save "the Knowledge Spheres" stolen by the invading Insidian Army.[74] However, these complicated narrative dynamics had little to do with the classic arcade-style action of the central game. Despite all the language that suggested a nuanced application of speech-act theory, such as the declaration that "the power to speak is the power to do," or even critical code studies, such as the assertion that "all resources are devoted to creating poetic codes,"[75] the literacy practices associated with the game remained relatively crude.

At MIT, a team of developers working with the Royal Shakespeare Company devoted thousands of man-hours to developing a Shakespeare-themed game based on *The Tempest*. As Squire and Jenkins explained the more sophisticated literacy practices expected of their players, "*Prospero's Island* is a space of dreams and magic, and students are encouraged to decipher symbols, manipulate language, and uncover secrets (in short, to perform literary analysis)."[76] Like 'Speare, there was no compulsion to set the action in Shakespeare's time or promote historical authenticity. Although some of the scenic elements in *Prospero's Island* were inspired by Renaissance curiosity cabinets and costumes that owed much to the fashions of Shakespeare's time, Squire and Jenkins were careful to emphasize the importance of staging Shakespeare in ways that were relevant to the contemporary world. They also situated the game-design process in the shifting discourses of literature departments: "There has been a significant movement in recent years away from conceiving the Shakespearian plays as sacred and unchanging texts, and toward studying Shakespeare as part of a living performance tradition."[77]

To Castronova's credit, he was remarkably open about the perceived shortcomings of the project and had the stated goal of allowing other game designers to learn from his mistakes. To understand what went wrong in his mind, it is perhaps useful to look at the program's initial objectives, which were spelled out in a FAQ.

> *Arden* serves many ends. People who play it will get to know the greatest writer in the English language without really trying. 2) *Arden* will serve as a test bed for research experiments, a Petri dish for social science. 3) By helping to build *Arden*, cadres of students will be preparing themselves for careers in the game industry and academia. 4) The construction and administration of *Arden* will create a locus of public sector expertise about the technology of synthetic worlds. 5) Finally, *Arden* will be a fun game, a good thing all by itself.[78]

Because Shakespeare has "battles, ghosts, dreams, elves, witches, drunks, sex, and pirates," Castronova believes "there's no reason it has to be boring." In the end, however, he declared Arden I to be a failure and "no fun" without the "puzzles and monsters" he loved. After the debacle, he began working on a less explicitly Shakespearian multiplayer game, *Arden II: London Burning*.[79]

Henry Jenkins has argued that, in their pursuit of fun, fan cultures are remarkably diverse in their interests and aims,[80] and certainly the continuing popularity of Shakespeare's plot, language, and characters demonstrates that the Bard continues to be a viable site for the kinds of fan behaviors characteristic of persistent games. Because of the possibilities for fun enabled by being able to choose genders, there are ways to dramatize how the heroes and heroines in Shakespeare's world experimented with social roles, as Marjorie Garber has pointed out in her work on the motif of cross-dressing in Elizabethan England.[81] It is strange that the "laboratory" that Castronova imagines didn't have much to do with this form of experimentation, particularly when online environments explicitly allow participants to experiment in the representational space.

Raph Koster has tried to put forward a unitary theory of fun based on brain science and puzzle-doing pleasures,[82] but the question remains, fun for whom? For the game designers? For the Shakespearian scholars who served as consultants? For the schoolchildren who beta-tested the game? For the teachers who had to give up classroom time for the exercise? Nothing draws attention to individual subjectivity quite like a theory of fun. Fun often appears to depend on each person's attitudes about pleasure, leisure, wish fulfillment, the bounds of Dionysian experience, or social practices of the carnivalesque.

Certainly, it didn't seem fun in the end for the game designers. Castronova said he made "some awful mistakes as a manager, which I don't hesitate to admit because, well, I am not a manager." The labor politics were not any better for those below him, whom he twice described as "slaves."[83] Castronova also assumed that it wasn't any fun for the Shakespeare scholars who worked on the project, who might have felt that the literature was trivialized or undervalued.

> Emphasizing Shakespeare was a mistake. The burdens of a license! Everyone thought it was World of Hamlet and the point was to teach high school kids 2B|~2B. But teaching Shakespeare has always been an ancillary benefit, not the point. I thought it would be cute. But putting Shakespeare in the game, I found, took away resources from fun. Lore, by itself, did not make a fun game. Shakespeare also loaded us up with an entire community of expectations, people who dig the idea of a digital Shakespeare.[84]

Lovers of Shakespeare were part of a user community that Castronova seemed to dread interacting with, one that he saw as "loaded up with expectations" and incapable of emergent play. Castronova perceived the texts of Shakespeare as ossified "lore" rather than a flexible rule set for interpretation and interpersonal interaction. Yet, it turned out that at least one of *Arden*'s Shakespeare scholars was still interested in the idea of embodied performance when I interviewed her. Professor Linda Charnes, who served on the project, subsequently participated in stagings of Shakespeare in

Second Life, and she expressed enthusiasm for the ways that *Arden* could allow participants to rethink what acting and action meant in Shakespeare's comedies and dramas.

At one point, while I was playing the game, long after the project was in deep hibernation, I left the keyboard and let my eleven-year-old son play as my character while I did some chores in another room. When I came back to resume play, he registered some disappointment at being kicked off the machine. "This is a fun game," he said to me. What's interesting to me about his summary judgment is that my child had no idea that *Arden* was an educational game. It wasn't differentiated in any way from the other games on my desktop at the time, which included commercial games like *BioShock* and *SimCity Societies*. I wonder if he would have felt differently had he known the game was intended to be educational. Perhaps that label inevitably dooms a game from the start. How would we view "educational novels" or "educational music" if we were told to consume them? How might the entire framing of the educational games movement be wrong?

Lost in Translation

In a course that Ian Bogost taught about adapting film and literary works for video-game genres, he suggested that thinking about "procedural translation" was essential in order to make thought-provoking independent games that improved the "current marketplace of Hollywood-bound blockbuster IP."[85] In other words, it was not enough to transfer characters, settings, and plots to video games, according to Bogost, if there was no serious consideration of the procedural features of the literary experience of the original work. As a teacher of game design, he pointed students to the Emily Dickinson "challenge" at the 2005 Game Developers Conference (GDC). *Gamasutra* explained the basic set-up, in which famous game designers sparred as follows: "In the 2005 Game Design Challenge, Will Wright returned to face off against Peter Molyneux and Clint Hocking in a challenge moderated by Eric Zimmerman. The theme? Design a game around a highly unusual 'license'—the poetry of Emily Dickinson."[86] As the introduction to the podcast of the panel explained, licensed material was conventionally thought of as derivative and thus inherently less interesting to game designers seeking creative opportunities. However, the GDC moderator also noted that this kind of intellectual property often had a great deal of market viability as well as an opportunity for invention.

Although he tried to do more than adapt Shakespeare's works to some of the existing genres of the educational video game, Castronova was disappointed when he wasn't able to realize a vast *World of Warcraft*–style multiplayer game with many different types of characters. Much of what Castronova said about Shakespeare in interviews and on his blog indicated that perhaps he didn't take this kind of translation seriously enough or engage with the philosophical possibilities of adapting Shakespeare for a video game, as a skilled stage director or filmmaker would have done. In fact, often Castronova sounded as if he would have rather adapted the works of fantasy novelist J. R. R. Tolkien for purposes of a video game, even though he said in a 2006 FAQ that "Shakespeare, as a source of game lore, is richer than Tolkien."[87]

In one of his postmortems on the project, he presented the following catalog of partial success: "Shakespearean quest lines; historically accurate tavern games; NPCs and resources drawn from Shakespeare; Shakespeare Q&A games that give experience points; Shakespeare text objects that grant power (text-as-treasure); Shakespeare texts accessed verbatim, in summary, and in quest/plot form."[88] Although this was a long catalog of game procedures, this inventory did not give much of a basic sense of the specific game mechanic or even indicate that he had worked out an effective overarching plot.

Some might argue that Castronova picked characters and settings for his Shakespeare game based on familiarity with a MMORPG player-versus-player combat aesthetic in an effort that was cursed from the start. By centering the story on the four Shakespeare plays featuring Falstaff and Prince Hal, one could argue that Castronova unconsciously repeated many of the same mistakes that almost proved to be the undoing of the filmmaker Orson Welles during the international production of *Chimes at Midnight*.[89] The opening scene of the game actually announces its status as a Shakespeare mash-up featuring an even more ambitious assortment of plays. The player converses with Peaseblossom, who explains some of the possible adventures that may unfold. The player can "talk to Falstaff to start a quest line involving Mistress Quickly," interact with the "rude mechanicals" of *A Midsummer Night's Dream*, "see a scene from *Love's Labour's Lost*," or start "on a quest that mimics the plot of *Richard III*." There are also references to other kinds of games, such as "Blackjack" and "Dungeons and Dragons."

By choosing the *Neverwinter Nights* game engine, the translation problem from page to stage to computer screen might have been made even more difficult. Thus, Castronova missed the opportunity to create a coherent computational work of art that mediated between real rules and fictional worlds[90] or exploited the matrix of binaries between diegetic and nondiegetic experiences, and the activities of operator and machine,[91] to foster experiences that might have drawn potential audience members to games and not literature in the first place. In addition to struggling with establishing how features of the general game genre would be integrated, Castronova also seemed to neglect what Bogost calls "special attention to the way specific aspects of human experience are represented in rules and code."[92]

It seemed as though the basic look and feel of *Neverwinter Nights* was much more radically redesigned when it was used in another educational game, MIT's *Revolution*, as a platform for exploring alternative histories and highlighting the practices of group deliberation that can be staged in game environments. In *Revolution*, colonists debate whether or not to support rebellious troops who are defying British authority, knowing that their decisions may have unanticipated consequences for members of different classes and professions. How *Neverwinter Nights* came to serve as the logical game engine for representing the historical past in educational games probably had its own history in the subculture of game development, although this digital "historicism" is obviously constraining.

By 2008, Castronova was emphasizing very different aspects of the *Arden* project from the ones that he had originally promoted to his MacArthur Foundation sponsors:

> Our experimental question (kept secret up to now) was: "Are fantasy game players economically 'normal'? Or on the contrary, when they make themselves into elves and dwarves and hobbits, do they stop taking economic decisions seriously?"[93]

The idea of keeping a central research question funded by a large philanthropic organization "secret" may sound unconventional to most academics, particularly those engaged in human-subject research requiring consent procedures approved by institutional boards of review, but Castronova maintained that his use of game theory and scenarios to model problems in *Arden* was a logical extension of the work of economists in his discipline. He eventually produced a paper, "A Test of the Law of Demand in a Virtual World: Exploring the Petri Dish Approach to Social Science," in which his team apparently "tested whether fantasy gamers conform to the Law of Demand" by manipulating the price of a given potion to see if "increasing the price of a good, all else equal, will reduce the quantity demanded."[94]

Alternate Reality Games

Elaborate multiplayer role-playing digital games can cost millions to develop, as *Arden* would have if the game had had the budget of the commercial games that Castronova knew best. So it is not surprising that more economically feasible games have had more sustained success in academia. In particular, alternate reality games (ARGs) have been able to engage large numbers of participants—often thousands of people simultaneously—with interactive narratives that use the real world as a platform. Such games may use transmedia storytelling techniques that deploy blogs, wikis, online videos, sound recordings, signage, comics, live-action role-playing, and improvisation. The "puppetmasters" who run these games aspire to create compelling and convincing experiences with intense player involvement by unfolding narratives that take place in real time and evolve according to participants' responses. Often, people are initiated into the game story after they encounter a "rabbit hole" element that encourages them to follow clues to uncover a larger mystery that demands their problem-solving attention.

ARGs are less resource-intensive because people generally play as themselves or some version of themselves, so digital avatars with programmed behaviors are not needed. For example, in the alternate reality game *Superstruct,* which McGonigal calls "the world's first massively multiplayer forecasting game" for grappling with five "superthreats" (disease, famine, struggles for energy dominance, privacy and civil rights crises, and population displacements), she makes first-person authenticity into a prime directive.

> 1. Play as yourself. Your character in this game is "2019 You." You don't have to use your real name, but please don't invent an entirely fictional persona for the game. After all, in the future, we'll all be some version of our real selves. So try to imagine your real self in the year 2019. And whenever possible, use your real life knowledge and real life strengths to help you contribute to Superstruct![95]

Second, these type of games avoid costly attention-getting effects, because they want to promote an immersive experience of play that blends game elements with supposedly normal life and to "deny or disguise" the fact that they are even games at all, so that nonparticipants may often be unaware that a game is underway nearby.[96]

Unfortunately, really good alternate reality games tend to introduce elements of unpredictability that campus staff might find unsettling. For example, when Annika Waern's team created the award-winning ARG *The Truth about Marika,* many people thought that its central missing-persons story reflected factual reporting and treated the case as a matter that merited contacting law enforcement. Campus-based games can similarly increase the workloads of college police forces when concerned citizens object. Unsavory elements may be created by either ARG designers or ARG participants, who are elaborating the story for themselves, so it can be difficult to enforce codes of conduct.

After the 2007 mass shootings at Virginia Tech, players of an ARG popular on college campuses found themselves subject to considerable scrutiny.

Humans vs. Zombies (or *HvZ*), which was created in 2005 at Goucher College, was banned at some schools because of its perceived violent character. Many saw the fake gore and pretend weapons as offensive to those traumatized by active shooter incidents, and some saw elements of the simulation as real. Reports about armed *HvZ* players at North Carolina State University put a campus that was already jittery after an incident at a nearby technical college into a state of full alert; on other campuses, *HvZ* participants were asked to sign legal forms that specified rules enforcing nonthreatening behavior.[97]

When alternate reality games are created specifically for colleges, they are often not very compelling. This is ironic, because many of the most interesting and critically acclaimed alternate reality games have been user-tested on college campuses, although they were not designed for them or by them. College campuses tend to have the right density of people in open public spaces, with enough potential players in the right demographic who already might be consumers of independent music or film and, therefore, might be open to participating in independent games. Yet many flawed game-design efforts, shown at conferences, have been created by misguided university administrators excited by the potential of ARGs to energize students but inexperienced in developing good games.

Often, college ARGs are made worse by campus planners trying to accomplish too many goals, such as combining the objectives of teaching students how to navigate the campus with increasing undergraduate fitness. For example, although Indiana University's *Skeleton Chase* used many of the common story elements of popular ARGs—including people reported missing and mysterious corporations—and benefited from access to interesting locales, such as a spooky campus greenhouse at Halloween, the real objective—to get students to walk at least 100,000 steps per week—took obvious precedence over the art of creating high-quality interactive fiction with meaningful networked participation.[98] Sometimes, campus games could even be quite inappropriate from a user-design standpoint, such as mystery ARGs set in libraries devoted to quiet study. Campus ARGs frequently demonstrate a mismatch between students' interests and those of administrators, as in the case of ARGs devoted to famous donors or little-known historical personages.

What Makes a "Good" Game?

Fortunately, there are some exceptions that prove the rule about the perils of investing in university-built games. The card-based deal-making game *Reality Ends Here*, which was organized for freshmen in the Cinema

School at the University of Southern California, generated an extraordinary amount of user-generated content from students and facilitated networking that persisted long after the game was over. (The game was named for the Latin motto of the school, "limnes regiones rerum.") Game cards might feature faculty members, film genres, and even industry labor unions as a way to encourage exploration and the creative combination of elements.

The researchers behind the game have described *Reality* in grand terms as "part of a broader educational reform agenda for college freshmen" for "connecting students to one another in ways that are both serendipitous and structured to maximize meaningful play and performance" in order to help transform "heavily siloed academic divisions into a productively chaotic and interdisciplinary community of practice."[99]

The term "community of practice" that *Reality* designers reference was developed by anthropologist Jean Lave and computer scientist Etienne Wenger to describe the apprenticeship processes of people in traditional trades, such as tailor, midwife, meat cutter, and quartermaster. What Lave and Wenger observed from watching novices learning the tricks of their trades turned out to be closer to what they called "legitimate peripheral participation" rather than the hierarchical and scripted labor relations associated with apprenticeship, because they noticed that these initiates often learned their trades experientially by observing, interacting, and copying rather than through explicit instruction from a master.[100] Peer learning also proved to play a significant role, particularly as learners took part in situated actions within a community of practice.

Wenger's "community of practice" concept has gained currency in recent years among researchers who study digital media and learning, particularly after being adopted by John Seely Brown, Douglas Thomas, Mimi Ito, Renee Hobbs, Henry Jenkins, and many notable others. However, there has also been a revival of interest in the traditional apprenticeship concept as a remedy for alienated teenagers who desire hands-on learning and recognition from adults.[101] By going back to Lave and Wenger's original research on groups of people who share a craft or profession, it becomes apparent why *Reality* designers saw this way of learning as appropriate for this large-scale heterogeneous "project creation game," given the fact that the auteurs-in-training at the Cinema School were expected to acquire considerable craft knowledge from specialized mentors in their field. The ambitious claim that such games might overcome the barriers of "heavily siloed academic divisions" more generally might have been more difficult to support, because these tight-knit students developed such strong affiliations to a single program.

Figure 8.2

Reality Ends Here website showing "deals" with game cards from USC. Courtesy of Jeff Watson.

Unlike other sponsored games in higher education, *Reality* was launched without fanfare. As members of the "Reality Committee" explained, participants "discovered it on their own, picking up on clues we've left around the campus—clues hidden in old cameras, left near our mysterious flag which intermittently hangs off the third floor balcony, or hanging from LED throwies we've stuck to the underside of staircases."[102]

The game is not mandatory for SCA students. It's not even openly publicized at the school. In fact, we've gone to lengths to try to keep it under the radar. ... Would we have had engaged all 180 of the freshman students if we had made the game mandatory? Certainly. But we doubt they would have been as committed, or that the game would mean as much to them. Students discover this game the same way they discover things like the college radio station.[103]

Curiosity was supposed to draw students into participation rather than extrinsic rewards, although it was precisely these rewards that were featured in later news coverage of the game: intimate meals with famous directors, private visits to television sets and studios, and other opportunities to hob-nob with celebrities.[104]

Of course, there were also many signs that the *Reality* project was clearly sanctioned by the university. The mysterious "game office" that students encountered may have cultivated elements of Kafkaesque intrigue, but the online presence of the organization at reality.usc.edu obviously indicated an approved, official domain that had the institution's blessing.[105] On the site, a short, humorous conduct video reminds students that they represent the university and cautions them about the hazards of guns, fire, rooftops, cars, the ocean, and trespassing in pursuit of the perfect camera shot.[106] The video also emphasizes the importance of being insured against liability with the motto "your project … your responsibility."[107] Unlike traditional codes of conduct that are likely to adopt a pose of paternalism, the USC video treats students as rational actors capable of avoiding bad decisions after a one-time warning. Students even felt comfortable enough about challenging the authority figures in the game office that they created a film parodying the competitiveness of the game.[108]

In lead designer Jeff Watson's dissertation, he contrasts *Reality Ends Here* with a comparable effort to engage game design students at the Rochester Institute of Technology: *Just Press Play*.[109] The slick *Just Press Play* video advertising the project boasts of their "collaboration with Microsoft," praises the contributions of "educational games," lauds the "gamification of education," and hypes a story based on "student heroes" and "academic dragons."[110] (In the video, we see the mood of a female student brighten after going into a male professor's office hours, as a game soundtrack signals the audience that she has triumphed over a daunting situation.) Particular kinds of interactions with faculty, such as finding a specific professor and making her laugh, get rewarded with prizes, such as cards with QR codes capable of unlocking achievements in the game.

For all of the granular specificity of such treasure-hunt activities in *Just Press Play*, Watson argues that these kinds of higher education games are designed to push for uniformity and standardization. Although a specific professor may be named for the make-the-professor-laugh task, she is actually a replaceable variable in the exercise who merely functions as Professor X from the standpoint of the game's designers.

Systems such as *Just Press Play* and the Open Badges project can be appealing to funding bodies because of their putative scalability and leveraging of computational

> automation. ... *Just Press Play* evokes a kind of "plug-and-play" or "set it and forget it" mentality that may be attractive to institutions looking to implement new programs within the constraints of lean budgets and limited personnel ... designing for massive scale is unlikely to be the most cost-effective approach to any kind of environmental design, as it results in flattened experiences that lack the fine-grain and high-touch attentiveness to the local raised by the emphasis on automation and surveillance that these and other such systems can exhibit. The degree to which participation can be rendered machine-readable is directly proportional to the degree to which that participation can be controlled and exploited. ... *Just Press Play* envisions a trade-off between the utility conferred upon learners and "far-reaching applications" with the "publishers and producers of curricular materials, and software and hardware vendors invested in the educational market."[111]

Watson alerts institutions to the risks of routinizing education by adopting gamification initiatives that allow data to be harvested. Much as Facebook collects data from their users that can be capitalized on by third parties, gamification efforts will be tempted to use the private information they have gathered from undergraduate players, as they inevitably seek ways to recoup initial investments in technology and design. According to Watson, initiatives like *Just Press Play* can therefore be much more pernicious than earlier educational games that were merely ineffectual.

Educational games have been justifiably mocked by game developers as "games not built by game designers," "chocolate-covered broccoli," and "games that are bad as games and bad as education." So what makes a good game in the context of higher education? Like many educators, I am persuaded that games can promote learning, just as novels, films, operas, and other media experiences can enrich human experience, deepen empathy, foster the exploration of different identity positions, encourage critical thinking, facilitate testing hypotheses, and help in deducing cultural rules. We should assign good games to freshmen, just as we assign good novels, but we shouldn't expect that they come with the "educational" label to indicate their wholesomeness.

Senior scholars have developed games to help them work out complex answers to research questions that allow them to consider many variables and outcomes. Many of these games have also been used in small-scale teaching experiments. For example, at Duke University, Tim Lenoir has created a game about conflict resolution,[112] and Katherine Hayles has designed one about financial speculation.[113] However, Hayles reported that when she recruited five proficient gamers to playtest *Speculation* as part of an independent study for course credit, the students played opportunistically to reach the goals of the game as quickly as possible and often missed important information about the history and ideology of speculative capitalism, which had been provided in a wealth of in-game texts in various media created by her team of designers. When the game was redesigned to emphasize particular opportunities for players to display virtuosity as thinkers and doers and to compete against each other rather than just to play against the game, the students got more out of the game experience, and the researchers were much more satisfied with the learning outcomes.

Such playable simulations differ from conventional computer models, because users can alter inputs to infer how different influences, catalysts, or factors may play a role in the outcomes that a given system generates and represents. In higher education, computational media have been developed for visualizing solutions to a number of problems in research areas that range from climate science to international relations. Like traditional role-playing games, students can study the interactions between different physical, biological, psychological, political, cultural, or national actors by experimenting with different combinations of inputs and forming hypotheses about causality. Such learning projects based on interacting with systems of rules certainly don't need to promote fun or happiness.

Procedural literacy events encourage students to figure out underlying rule sets through experience rather than through didactic delivery and direct instruction. By experimenting with different algorithms at work in a digital representation, students can see how a set of implicit rules can be made explicit. For example, using computer programs that are created by researchers in artificial life, students can see how patterns of segregation can unfold in urban neighborhoods.[114] Much as "oral performance surrounding a written piece of material" can be a key ritual of membership in a community, as described by Shirley Brice Heath in her work on "literacy events,"[115] students performing at a procedural literacy occasion must deduce the rule at work in a given system and then verbalize how it operates in procedural literacy events that solidify community membership. Those good at decoding (or encoding) information are subsequently recognized for their achievements with enhanced forms of social membership. Procedural literacy can also be cultivated using collective-intelligence activities through cooperative group work in which groups are allowed to puzzle-solve together to articulate the rules.

Most of all, such games, simulations, and virtual worlds don't require expensive investments in digital technology to be effective. The recent success of the Polish card game *Kolejka*—"Queue" in English—about pre-Solidarity life behind the Communist Iron Curtain demonstrates that history can be taught quite effectively to a variety of learners in very low-tech ways.[116] It may be obvious that enthusiasm for learning with 3D avatars in Second Life and *Arden* seems dated now, but educators should also be wary of other headline-grabbing trends. Although badges and gamification schemes can be effected with relatively little investment and minimal technological infrastructure, they serve our students poorly and do not help them develop the agency they will need to solve the real-world problems that we face.

9 Gaining Ground in the Digital University

The students did not appear to be immanently in danger of being naked when I arrived, although one of the male students arguing with the authorities was already shirtless. Given the campus's proximity to the beach, seeing scantily clad undergraduates in bathing gear—as these freshmen were—was relatively common. Furthermore, the university had long been known for incorporating a certain amount of West Coast hedonism into its student culture. An annual "undie run" held during spring final examinations encouraged hundreds to dash through the library wearing only their skivvies.

Yet a dean and an assistant dean of residential life had wanted to prevent the planned "naked water balloon fight" that had been advertised through a Facebook group and for which 110 enthusiastic students had already RSVPed.[1] When I got there, the officials from student affairs had a camera on a tripod prepared to capture any conduct violations. They were sternly discussing possible legal consequences and the likelihood that emergency first-responders could soon become involved if the students weren't willing to relent.

I was present for this standoff between the professionally clothed and the partially unclothed because the rationale for potential nudity was an assigned class project. As part of a capstone experience in project-based learning, first-year students were required to collaborate in groups and then to exhibit their research or creative findings in presentations, performances, or poster sessions at the end of the year. This particular group of students, enrolled in a course taught by anthropologist Joe Hankins, was supposed to conduct an "ethnographic intervention" that involved disrupting a pattern that they had observed, and the students—inspired by readings from Kate Bornstein's *My Gender Workbook*[2]—had chosen to disrupt common conventions around modesty and avoiding the display of gendered sexual bodies in public.

When I first got there, I had assumed that the students in the confrontation were those of famed hacktivist professor Ricardo Dominguez, who was teaching in our program and who was known for promoting electronic civil disobedience and for risking loss of his tenure as a result of his use of tactical media in digital protests against corporate capitalism, Western political hegemony, U.S. immigration policy, and tuition hikes at public universities.[3] I knew that one of Dominguez's lectures had discussed transgender women protesting topless, and I had already met with another college dean about another Facebook-coordinated class event for his class. Dominguez's students had created a fake naming ceremony using the college's official logo to announce that Sixth College would henceforth be known as Nikola Tesla College.[4] The naming-event organizers had promised to distribute free consumer electronics to a Facebook group of over four hundred people who apparently planned to attend, and campus authorities were worried that the disappointed crowd for this nonpermitted, nonstaffed event could degenerate into a group of undisciplined rioters.

Not only were there the would-be naked water balloon fighters and the Tesla College pranksters to contend with; other groups had been scheming to get attention by putting up caution tape to interfere with routine jaywalking, disabling wireless routers, covering walkways with temporary graffiti, and disrupting use of the gym. The college's staff had understandably become exasperated with running around to forestall chaos from multiple unsanctioned, unpredictable student activities, all designed to propagate emergent behaviors and cascading effects. I had to serve as a mediator between the students and the administrators by arguing that legitimate political resistance could take the form of satire and that student engagement in coursework was definitely demonstrated by the enthusiastic nature of their participation. At the same time, I wanted to show respect for institutional procedures, financial resources, and the time of the college's staff, many of whom already resented faculty governance structures that treated them as members of a permanent underclass, despite their professionalism and years of service.

A heated discussion took place, in which one college official loudly complained about what he saw as the elitism of a professor who could hide behind academic freedom for his own actions but was risking expensive and risky disciplinary consequences for his students. He also accused the professor of not caring about the labor of nonfaculty personnel responsible for campus safety and student well-being. He ridiculed what he saw as the triviality of the bourgeois goals of the protest in comparison with preventing gross forms of gendered injustice and violence, and he suggested that the students should devote their activist energies to ending clitorectomies in Africa instead. Fortunately, the instructor of record and the deans eventually talked via cell phone and scheduled a time to meet face-to-face to discuss future cocurricular collaboration strategies. After almost an hour of dialogue, everyone dispersed peacefully without any illegal activity occurring.

If Clark Kerr was correct in asserting that the university really is a "multiversity" in which students, faculty, staff, alumni, and many other groups must coexist and cooperate,[5] conflicts like the one that unfolded in that grassy quad, with the water balloon demonstrators who planned to strip naked, will always have to be negotiated with sensitivity to all parties. Of course, it requires human listeners to sort out the claims of all of these voices and to facilitate meaningful dialogue between them.

I tell this story because I worry that if the university of the future becomes largely an online entity, making sense of such conflicts between stakeholders will become much more difficult. All of the possible interpretations of the student behavior—as protest, as prank, as vacuous exhibitionism—could apply to user behavior in environments of computer-mediated communication. However, it becomes much harder for all viewpoints to be heard, and for any kind of consensus to be reached, if a remote webmaster ultimately controls the user experience. If student codes of conduct become similar to impersonal and legalistic end-user license agreements, such as the click-through agreements for Facebook or iTunes, what happens to the possibilities for articulating problems generated by fundamentally inequitable relations of power? In an online university, a disruptive student can be instantly silenced by being disenrolled, and a bot can delete any mentions of potentially offensive terms like "naked," along with the posting person's access to communication with faculty and fellow students.

As Lawrence Lessig has observed, technical architectures can prohibit many types of free speech and civic discourse.[6] In the university of the future, it may be difficult to have civil discussions about controversial readings that may contain automatically flagged terms such as "house nigger" or "office bitch." Much as many forms of legitimate, critical discourse in remixed digital files may be prohibited by pattern-matching algorithms on corporate video-sharing sites that police the use of intellectual property or offensive content automatically with dumb technologies, we need to think about how online educational spaces will be regulated if institutions scale up to accommodate participation by thousands of enrollees and let the oversight be handled by machines.

Machines seem to have vast powers of surveillance, but the problem is that they can't really listen. This point about technology's limited powers of perception is emphasized in the work of anthropologist Lucy Suchman, who adopts the viewpoint of the machine as a way to understand human–computer interaction and analyze frequent breakdowns in communication. In 1979, Suchman was hired by Xerox PARC to conduct ethnographic fieldwork. She soon realized that users often overestimated their devices' abilities and treated them as more intelligent agents than the machines really were. As Suchman asserts, "the machine could only 'perceive' that small subset of the users' actions that actually changed its state … as if the machine were tracking the user's actions through a very small keyhole and then mapping what it saw back onto a prespecified template of possible interpretations."[7]

Suchman also recounts a well-known anecdote about the differences between a European navigator, who traverses the open ocean with well-articulated, advanced planning based on universal principles, and one from Micronesia, who knows the geographical goal but not the specific course and responds to particular circumstances with embodied knowledge. Using this story, she argues that all of us actually behave like the Micronesian navigator as we adapt to contingencies, and we use the planning model only afterward to justify our behavior in the moment. Her theory of situated actions, which she developed watching people interact with machines, can be a useful way to approach our enthusiasm about adopting particular policies toward instructional technology. In rapidly changing teaching environments, like the navigator, we need to stay nimble because we can't always see the entire trajectory that will lead us to the future.

This need for a flexible vision brings me to the central issue of this book: How can we influence the digital university to be more inclusive, generative, just, and constructive? How do we learn from embodied interactions in the lived experiences of the current university's users, rather than simply accepting the untested structures of entrepreneurial venture capitalism—structures that might disenfranchise many kinds of learning activities when such activities fall on the wrong side of a cost–benefit analysis? How do we avoid overestimating machines? Based on my two decades in the trenches, I suggest six relatively simple principles that could guide effective pedagogy and decision making.

(1) The Golden Rule Should Dictate Decisions about Instructional Technology

One of the realities of being a faculty member in the contemporary world is that one never stops being a student. Going to lectures, reviewing the work of others, and being tested on knowledge—at least informally—will likely continue to be part of the robust professional career of any active scholar, even one who attains tenure or settles into an administrative position. This is why it is mystifying that faculty embrace knowledge-sharing methods that they themselves would find highly intrusive if student-teacher roles were reversed. Showing attentiveness, answering questions directly, and being honest about having only incomplete information are appropriate behaviors whether one is an instructor or a learner. In contrast, assaulting others with digital countdown quizzes or spying on their social network profiles are aggressive tactics that professors certainly wouldn't want used against them if the tables were turned.

As though aspiring to read students' minds, some pedagogical initiatives even monitor involuntary responses. For example, the Gates Foundation announced plans to invest 1.4 million dollars in having middle-school students wear biometric wrist monitors that measure changes in the sympathetic nervous system, supposedly as students respond to stimuli generated by the instructor. Described as "engagement pedometers," the wristbands are intended to record when material is exciting to students and when it falls flat. Rather than video-recording classroom interactions and then having a human expert analyze the footage, or documenting how students participate with paper-and-pencil methods,[8] school administrators are promised "universal, valid, reliable and practical instruments."[9] The devices, known as Q Sensors, are produced by Affectiva, an MIT Media Lab corporate spin-off, which was cofounded by affective-computing expert Rosalind Picard. The company is also known for developing facial-interpretation software that helps advertisers gauge the success of product campaigns.

Such intrusive technologies are easier to implement in college classrooms, where students are likely to be adults, because testing hypotheses with minor children often requires parental consent. In addition, faculty in university settings can often avoid completing paperwork for institutional review boards governing research on human subjects because undergraduate pedagogy is frequently an exempted category. Even though resentment and negative reactions are common responses among subjects who are experimented on like guinea pigs, instructors might be eager to pursue grant money that rewards risk taking and its associated prestige. Troubling aspects of the so-called "quantified self" movement are often ignored by these true believers,[10] even though so-called "persuasive technologies" can easily be as constraining as overt architectures of control.

Part of the problem of applying the golden rule has to do with defining the term "student engagement." Does engagement describe the external activity a student demonstrates during the course of a class period? Or should it account for how much internal reflection is happening inside the student's head? Does long-term retention of information many years later matter? Or does the application of course knowledge to problems in real life count most? Different types of students engage differently at different time periods in different modalities. An art student doodling during a lecture might signify intensive engagement with material, while drawing in class might have an entirely different meaning for a daydreaming biology major. Even students poring over the assigned book or writing drafts of a required paper as catch-up during a lecture might be called highly engaged, even if the type of engagement is not appropriately scheduled. Too often "student engagement" is code for "faculty disengagement," particularly when the computer serves as a substitute for the professor.

Applying the golden rule doesn't mean that students should be deprived of the security of knowing that someone human is in charge. Totally flattened hierarchies can undermine confidence that the instructor is a reliable expert, and time-conscious students may prefer listening to the most knowledgeable speaker in the room rather than the most talkative one. Too often, instructors abdicate their authority in the classroom and put the technology in charge, which shortchanges students who have come to the digital university to learn. [11]

(2) Faculty and Students Should Use the Same Tools

Although laboratory equipment for specialized experiments may be too expensive for undergraduates to handle, in most cases the academic technologies and research tools that faculty already use can be made much more accessible to their students. Promoting student access to existing digital resources on campus—such as those for analysis, mapping, imaging, and visualization—is far more important than purchasing new equipment with limited uses. Expensive clickers, iPads, licenses for learning-management systems, or other costly instructional technologies that are only used by students do little to support scholarly endeavors, and they also reinforce the notion that students are second-class citizens who only deserve non-faculty gadgets. As the Boyer Commission has observed, undergraduates at research universities often have little contact with the process of discovery that is ongoing at labs, clinics, studios, archives, and research centers,[12] and instructional technology does little to provide the kinds of experiential learning or completion of capstone projects that have been identified as high-impact educational practices, such as those recommended by the American Association of Colleges and Universities.[13]

Successful digitally enhanced classes have used HIPerWalls (high-resolution interactive parallelized display walls), caves, visualization portals, virtual worlds, participatory screen systems, information visualization software, geographic information systems, rich media publishing systems, teleconferencing, 3D modeling labs and light stages, digital editing bays, machinima, video games, robotics, 3D printers, milling machines, and even paper prototyping. These technologies usually are not purchased for pedagogical reasons; they were budgeted to support faculty research. Like archival special collections in rare-book rooms, research spaces are often seen as off-limits to undergraduates. Yet students can manage complex tasks for faculty with adequate training, and they gain much more from learning to use scanning electron microscopy machines, 3D databases of ancient artifacts, or video annotation software than they learn from clickers or iPads that emphasize short-term skill-and-drill retention. For example, at the University of California San Diego, undergraduate students working with media theorist Lev Manovich were encouraged to develop projects using large data sets from art, literature, design, video games, fashion, and music. To provide appropriately detailed media visualizations, which might include thousands of paintings or buildings or design portfolios, students viewed their work on the campus's HIPerSpace Wall, which offered one of the world's largest displays with screen resolution up to 220 million pixels. Students in the class essentially composed giant posters for their final projects that were uploaded for critique sessions and public viewing. Ironically, the building in which these class meetings took place generally prohibits such pedagogical uses of its rooms. Yet it is precisely when these "off-label" uses of digital technology happen, when scholarly resources can be appropriated by students to deepen hands-on learning, that students develop a strong sense of their own agency. Although research universities are likely to have more equipment, they unfortunately also tend to have more barriers to pedagogical uses than community colleges, liberal arts institutions, and design schools with strong teaching missions.

This tendency to foster a separation between the tools of faculty and the tools of students can cut both ways. Faculty members tend to deny the fact that their own research methodologies using digital collections might rely on common consumer search engines, such as Google.[14] They may tell students to shun sources found through Google search, Amazon's "look inside this book" feature, Wikipedia, and other commercial sites, but social scientists studying academic practices over the shoulders of faculty members in

Figure 9.1
Interdisciplinary Computing in the Arts students with the HIPerWall. Courtesy of Lev Manovich.

front of keyboards have noticed that professors often undertake research and writing projects by pursuing, at least initially, many of the same routes as their students.

Classroom spaces that allow students to share their computer screens in situations where all parties are using similar tools can encourage a sensibility that supports the development of a joint intellectual culture. For example, in the Zemeckis Media Lab (ZML) multiscreen space at the University of Southern California, which was designed by educational technologist Scott Fisher, instructors encourage students to contribute information from their individual computer screens to the group. As students contribute resources to the communal display space, the instructor is likely to encounter images, links, additional commentary, and other relevant sources that would have been otherwise overlooked. In this way the paradigm of what Pierre Lévy has called "collective intelligence"[15] can supplant the much more problematic surveillance paradigm and can imbue students with a much stronger sense of the purpose of group interactions.

(3) "Old" Technologies Should Still Matter

In an essay called "A Pedagogy of Original Synners," Steve Anderson and Anne Balsamo, who have taught in the ZML space, describe how they might manage a classroom in 2020 in which learners are instructed to "Pick your Medium: Physical, Mental, Chance, or Arts" and then choose among following options: "a) naked, b) tool, c) machine, d) animal." The third option, "machine," emphasizes futuristic high-tech "digital devices and applications, as well as engines, robots, biolution devices, flickercladding and other nano manufacturing gadgets"[16] that are imagined as used collectively by instructors and students doing the subversive "citizen science" work of what Beatriz da Costa and Kavita Philip have called "tactical biopolitics."[17] However, the second option, "tool," includes many present-day, low-tech aids to discovery, such as "markers, dice, picks, hammers and pens,"[18] which are already in use in prototyping, especially in rapid paper prototyping for video game development or digital interface design.[19]

Anderson and Balsamo's suggestion that low-tech instructional technologies deserve more attention echoes the advice from other digital educators that "old" media should never be ignored. For example, digital rhetoric specialist Cheryl Ball recommends "paper, crayons, scissors, tape, the Web, their smartphones"[20] as essential tools for teaching computational media, and critical making guru Garnet Hertz, who teaches technically sophisticated courses on robotics, programming, and video game design, celebrates the virtues of typewriters, zines, and paper prototyping as tools for thinking.[21]

Great pedagogy can involve engaging large classes in print publication activities and refusing to join uncritical celebrations of a future without books. For example, New School faculty member Trebor Scholz asks students to submit print-on-demand publications that consist of at least ten thousand words in lieu of a traditional final exam for his new media courses. According to Scholz, students appreciate having produced an actual book as a physical object at the conclusion of the term and are inspired by their "pride of authorship."[22] The students of the University of California, San Diego Ethnic Studies professor Wayne Yang host their own Comic-Con comic book convention at the end of the term, where they curate original, research-based graphic novel projects using library sources. Although print artifacts are central to the culminating activities of the course, students also work on their digital skills as they lay out pages in software programs or post video trailers on YouTube. They also explore other "old" media by producing convention swag and, thus, investigate the semiotic properties of t-shirts, buttons, signage, and other designed objects. Yang's work with comic books was influenced by the work of Ernest Morrell on critical literacy and urban pedagogy[23] to promote "a taxonomy of different kinds of literacy," including "traditional media literacy" and "new media literacy."[24]

Creating effective exercises in critical making that involve atoms rather than bits does, obviously, require thoughtful planning and administrative support. One-of-a-kind student work that uses paper and other tangible media can be surprisingly resource-intensive to produce, display, submit for credit, mark with feedback, return to students, or store. It is for this reason that studio art departments enforce draconian policies to avoid filling up all their available spaces with student work. Even if the focus in a given course is on the pedagogical event rather than the tangible artifact, it can be difficult to negotiate the territory between treating student work as though each piece is a precious institutional treasure meriting preservation for posterity and dismissing such output as disposable arts and crafts. Different learning objectives also shape policy: book production activities emphasize

Figure 9.2
Sixth College Comic-Con capstone print publication activity. Courtesy of Andrew Mandinach.

the *product* generated as evidence to be evaluated, and paper prototyping exercises focus attention on the *process* learned.

(4) The Occasions Should Be Joyful

Current theories about the triumph of telepresence and ubiquity often assume that the digital classroom will eventually become obsolete. When rich media experiences optimize bandwidth efficiencies, when devices become inexpensive and portable, and when all human knowledge is digitized, then learners—the logic goes—can be finally freed from the constraints of the classroom. Book groups, meet-ups, and flashmobs in improvised physical spaces will satisfy our ancestral needs for face-to-face interaction as primates, and the activities once devoted to study, sociality, and activism within campus geographies and architectures can be disbanded. As Stephen Ramsay has reminded his fellow college professors, the classroom is an eighteenth-century historical invention that can be traced back to Pietist educational reformer August Hermann Francke; he "put forth the radical idea that there was a time for listening and a time for speaking—a time for being educated, and a time when that process would end."[25] Just as there was a time before classrooms, there may be a time coming after them.

Traditionalists who fear the advent of a brave new world might wish to take refuge in grumbling about the loss of memory, attention, and privacy associated with institutional culture and on-site campus work, but even better arguments for reviving the classroom can be made. Learning benefits from embodied performances of knowledge that respond to the exigencies of specific rhetorical situations occurring in time and space. In other words, classrooms that incorporate special occasions can make education feel more special, but repetitive skill-and-drill exercises that emphasize tedium and task-oriented monotony can make one class session feel just like another. Coordinating hackathons, open-mouse nights, game jams, Wikistorms, and other one-time events can provide intense experiences of presence that make "live" digital instruction much more meaningful by enhancing solidarity and reputation in inhabited affinity spaces.

Within software culture, a hackathon is usually a day-long or week-long event during which computer programmers and software developers collaborate intensively on projects, usually face-to-face. In academic contexts, hackathons offer opportunities for live performances of knowledge in technologically enhanced spaces. For example, MIT's Nick Montfort frequently uses the hackathon format with his students. Rather than have students iterate in isolation and privacy, writing lines of code becomes part of a communal ritual of participation. In this frenzied environment of rapid and uninterrupted output, his students create work that can be shared and curated online, in gallery spaces, and in a variety of other public settings. In recalling his own undergraduate experiences with large, impersonal lectures, Montfort claims that drafting and teamwork are the most important skills to be learned in college—and yet students mostly gain competence in these areas *outside* of the classroom.[26] In the hackathon environment, such composing, revision, and collaborative activities become central. Hackathons are not limited to work by students either: Roger Whitson has used the hackathon to generate digital humanities curricula,[27] and Dave Lester has used it to mash-up syllabi.[28] Although Lilly Irani has justifiably raised questions about the hackathon as a design milieu in which a code of machismo and neoliberalism dominates,[29] competitive live coding events that recognize achievements of speed, endurance, power, ingenuity, economy, or aesthetic finish can be inclusive as well as motivating. If properly imagined, these exercises in rapid prototyping can accommodate many kinds of learners, modes of collaboration, articulations of identity positions, and areas of expertise.

Another type of digital pedagogical occasion might emphasize solo performance. For example, Mark Marino has worked to perfect what he calls "the open mouse night" as an opportunity for students to display their prowess as digital media makers, much as writers who create work for print media gather to share new work, old work, and works in progress in front of a microphone. Borrowing from the conventions of informality and acceptance around open mike nights, Marino urges everyone to participate in an environment of mutual celebration. Because many digital works are nonlinear and have multiple paths for navigation, audience participation often adds to the excitement of these events.

Feminist and antiracist "Wikistorms" gather digital authors together for periods of intensive activity to edit the massive online reference work Wikipedia. These events focus on documenting the cultural contributions of underrepresented groups and converting more materials from scholarly print archives about women and people of color to the Web. Specialized journal databases of licensed content behind paywalls for institutional subscribers make information available to scholars and students, but Internet searches by the general public only can access material that is free. Wikistorming brings knowledge from campus libraries and translates it to the Web.

As Wikistorming facilitator Adrianne Wadewitz explains, students can be important contributors to Wikipedia. Wikistorming can serve as a consciousness-raising exercise, because "the point of doing feminist outreach is you need to find not only women but also feminists." In considering who gets "written out of history," she wants to encourage active questioning of "the structures of knowledge" and promote understanding that "every edit is political," because "you can't put in every piece of information and you can't use every source."[30]

For example, in a 2013 class session orchestrated by Wadewitz at Pitzer College, students focused on edits that would enhance the accuracy of Wikipedia's information about women and technology. Under her tutelage, students reviewed topics that ranged from "Afrofuturism" to "Women in Video Games." During this compressed time period of focused activity, the undergraduates improved the user-generated content in the database by providing sources and rewording language on entries for a range of feminist scholars, including Karen Barad, Brenda Laurel, and Faith Wilding.[31] However, students also learned important lessons about limitations to publishing material with original research and their appropriate roles as apprentice editors on the site. In a sample training video that shows Wadewitz interacting with student editors, she emphasizes the importance of favoring "verifiability" over "truth" and understanding the community dynamics among Wikipedia editors.[32]

Figure 9.3
Wikistorming activity with Alexandra Juhasz's students at Pitzer College. Courtesy of A. J. Strout.

(5) The Issues Should Be Serious

The Wikistorming exercise is intended to have real-world consequences and to reshape an important venue for public knowledge. Students are urged to take the activity seriously and to grapple with substantive ethical questions. Unfortunately, too much digital instruction comes from well-meaning, enthusiastic faculty with little more than a vague objective to empower students. In particular, unstructured make-your-own-remix exercises can generate student work that is superficial, uncritical, amateurish, or naïve. Although scholarship from Lawrence Lessig and other critics about remixing as a political tool can be valuable in guiding good pedagogy, students don't always have the critical capacity to interpret originals, much less derivative works. For example, a typical prompt asks students to "create a remix video that in some way offered a critical examination of media, posting it to YouTube to potentially generate some feedback from people who stumble across it."[33] Such an assignment gives students little guidance on how to locate an audience, apply shared critical frameworks within the broad field of media studies, or see how and why remix makes sense as a rhetorical strategy. After having looked at thousands of student remix videos posted on the Web for assigned work, my radical suggestion is that we need more *unmix* pedagogy.

Labeling primary sources accurately can be a difficult task. Professional librarians curate little of the material that is uploaded to the Web. Metadata conventions are usually ignored. This creates a wonderful opportunity for students to work with remixed video from news programs, advertisements, artworks, and political commentary and perform unmix activities. Given a mash-up video drawn from multiple sources, students could be asked to locate precisely where each section was shot, under what circumstances, when, for what purpose, and by whom. A variety of free and open source tools on the Web can make it easier to identify digital signatures of video and audio sources. Shot detection and facial recognition software can isolate discrete components. Even commercial products such as Shazam or Google image search can help source the provenance of individual shots.

Students can also use crowdsourcing to supplement their own knowledge bases about the contents of remixed digital collections with input from others who might be more expert in identifying images, footage, soundtracks, or texts. Much as traditional classrooms teach forms of explication, close reading, word study, and lexical analysis that allow scholars of print culture to trace significant patterns of appropriation and reuse, digital classrooms can teach rigorous interpretation of digital compositions.

Figure 9.4
Example of shots from Al Jazeera news that could be unmixed by students. Courtesy of Jeremy Douglass.

Being serious doesn't mean being intolerant of subversive influences. Satire can be a legitimate mode of intellectual discourse, even if most student "parodies" demonstrate little knowledge of the originals. Such "dark values" in design can be useful to acknowledge as a way to promote critical thinking,[34] and even recognizing a role for "evil media" within digital culture can be useful.[35] Seriousness should not be confused with priggishness or moralism, since devil's advocacy and provocative conversations are an important part of the academic experience as well.

Unfortunately, online learning often takes the sharp edge off of what works best in traditional higher education, where biases of students are challenged, and comfortable generalizations must be tested. The major role of philanthropy in many educational initiatives—often in the name of good corporate PR—probably further reinforces an aversion to controversy in instructional technology efforts, along with a tendency to divert attention from difficult questions about diversity to focus on photo-opportunity tokenism instead. Moreover, critical pedagogies tend to be watered down when people must come to agreement from very different camps, in this case from clinical behaviorism, business entrepreneurship, and sixties anti-capitalism. Uttering the same slogans in the name of educational reform pushes a culture of uniformity and enforced consensus and stifles the kind of dissensus that college campuses nourish.[36]

(6) The Novelty Should Have Worn Off

The worst reason to implement a new instructional technology is because it is new. Yet newness is the most common reason that educational gizmos get adopted in the first place, even though the assumption that students love novelty is almost always wrong. The folly of overvaluing innovation is one of the major themes of this book. MOOCs, badge systems, gamification, iPad distributions, and Second Life grab headlines by appealing to trends. I predict that they all may go the way of the magic lantern by the time the next century rolls around.

Students need models that have been tested, not fads. When asked to submit work or respond to a question, they need specific information about audience, purpose, format, and genre. Clear direction and concrete suggestions about appropriate process have long been recognized as valuable for students grappling with traditional assignments, although they also need to be challenged with open-ended problem solving activities to transfer knowledge to new domains. Faced with a vacuum in instruction, students improvise, which can stimulate expression just as likely to be banal as it is to be liberating—and behavior that is just as likely to be unethical as it is creative.

It is not difficult to use available technologies to improve instruction. For example, e-mail is often undervalued as a pedagogical tool. When I teach large lecture courses, I find electronic correspondence to be one of the best ways to get to know my students and for them to get to know me. I read their personal statements for graduate school, I click the links that they think I might find interesting, and I even download music they suggest as appropriate for the course.[37] After decades of writing e-mail, I can compose quickly and confidently and model good compositional practices for my students. With brand new technologies I may fumble.

Student evaluations can be another tool for improving pedagogy. Faculty usually don't enjoy making numerical course evaluations available on the Web, but posting the results of official comprehensive evaluations makes students less likely to rely on idiosyncratic "reviews" with arbitrary scoring systems, such as RateMyProfessors.[38] This is an extremely cost-effective way to improve attention to teaching, because it involves the public reputation of an institution's faculty.

Unfortunately, a myopic focus on the newest digital technologies can make it difficult to think about instructional technology as a broader category. Windows in a classroom that bring in natural light are an instructional technology. Chairs that can move, so that students can see other student faces, are a technology. Instructional technology needs can be fulfilled just with adequate whiteboard space and a sufficient number of dry erase markers for listing recapped lecture material or letting students display their own claims on the board for purposes of comparison.

Sometimes we can learn more about instructional technology by studying technologies that aren't explicitly labeled as instructional. Good teachers often think broadly about learning, and they train themselves to pay attention to learning activities taking place everywhere: at home, on the street, and on the Web. Large numbers of users are already adapting their behavior in response to adopting commercial devices and platforms, and these new forms of conduct have inspired a significant scholarly literature. Publications in the narrower ed-tech field may contain fewer useful insights by comparison. Important research also comes from all around the world; it examines a diversity of practices that includes mobile money transfers, social networking among migrants, witness journalism, protesting flash mobs, online scams, and computer hacking. All of these populations are learning as they interact with new technologies. How can we apply these lessons to the classroom?

However, success in implementing a new instructional technology depends on having a critical mass of faculty colleagues willing to share their experiences. Such pedagogical communities require there to be enough trust that failures can be assessed honestly and lessons learned can be shared without jealousy and covetousness. Often, what AnnaLee Saxenian has called "regional advantage" is also an important factor.[39] Clusters of colleges connected by transportation corridors facilitate enough periodic face-to-face contact among people from different institutions to develop networks and exploit enclaves.[40] Collectives of experimenters are needed so that projects can flourish after the first giddy launch period or—if appropriate—be allowed to die gracefully. By using hybrid instruction rather than pure distance models, lots of data can be gathered as well.

Two-way communication between instructors and students is another necessary condition. For effective digital pedagogy to function, learner-participants must be allowed to air concerns about access, equity, usability, and sustainability and to raise objections to proprietary software, costly hardware, or untested prototypes. The channel for communication should express the values of the institution, and thus commercial services can be problematic. For example, Facebook may be a good way for students to communicate with each other, and it may even be a good way for informal intrafaculty contact to share news. But Facebook is notoriously poorly designed for faculty-student exchanges. Even if the contact is student-initiated, Facebook friending risks forms of social surveillance from either party that are probably inappropriate, and it deepens a confusion about already confusing roles.[41] The reposting of Facebook "fails," on the popular site FAILBlog, involving embarrassing teacher-student Facebook contacts indicates the dangers of such blurring of hierarchical boundaries.[42] Not surprisingly, many students already have worked out protocols that involve maintaining multiple online identities for dealing with their elders,[43] or they rely on sites that perform less affective labor, such as LinkedIn.[44] Nonetheless, corporate social network sites remain a considerably less than optimal choice for academic communication. After all, choices about code, platforms, and infrastructures express particular values. Unfortunately, campus-specific or course-specific social network sites designed for educational uses are also problematic choices; academic social network sites tend to be underused by students, time-consuming for faculty to set up, and frustrating for support staff who want integration with learning management systems and reliable products that won't be discontinued or likely to change their pricing structures abruptly. For example, when users of the community-based social network service Ning were told that free services were being discontinued, many educators were disappointed to see that use of the software was not sustainable.[45]

There also needs to be effective communication between students in peer groups to develop robust digital pedagogies. Students pursuing course optimization solo, who are devoted exclusively to autonomous leveling-up, suffer from what James Paul Gee has called a "school of one" mentality that makes it difficult for them to contribute much to the larger social entities that sustain higher learning institutions.[46] When inter-student communication is successful, there is much that can be learned about successful strategies to improve teaching, with or without technology. For example, a group of students in James Fowler's social networks course at the University of California, San Diego studied how the instructor's position in a large lecture hall influenced the likelihood that students would speak up and contribute during class. Although aspects of their initial hypothesis, that students in the front row would be more likely to talk than those in the back row, seemed plausible, they observed a more complex pattern of "contagion" in which clusters of students all around the lecture hall participated verbally and did so in response to the presence of either the lecturer or an active student-participator.[47] In this way, a talking body activated nearby students to talk. Although these student researchers used a computer laptop to show some of their observation results, they also created an analog information visualization with an eye-catching do-it-yourself aesthetic that expressed the craft community of their peer group.

Figure 9.5
Sixth College students researching lecture hall participation showing their results. Courtesy of Eliza Slavet.

Much of this book is devoted to failure stories: embarrassing professorial breakdowns captured on video, cynical students cheating with digital tools, curricular meltdowns when course management systems become overloaded, false prophets of instructional technology pushing geegaws that will soon be derided as worthless, and many other examples of mistakes and miscommunication around attempts to manage the unstable connections between formal and informal learning. I often begin my analysis with a story of failure. I start off chapters by describing pedagogical events in which everything seems to go horribly wrong.

The larger story is a more complex, more hopeful one, which is grounded in the optimism of having chosen teaching and learning as an avocation. It is a story about using technologies to make sense of the world and of each other and a story about better appreciating the imperfect character of our knowledge as a result. The "learning sciences" movement is fixated on achieving perfection with a single tool, technology, or platform, even though the *real* sciences understand that universities, learned societies, and bodies of scholarly knowledge that have developed over the course of centuries make progress toward finding solutions in much more uneven and unpredictable ways.

One of the ironies of the learning sciences movement is that its claims are frequently not scientific. As I have pointed out in the course of this book, exhaustive research from both ethnographic fieldwork focusing on specific case studies and large-scale quantitative experimentation in the social sciences indicates that many of the premises of the instructional technology movement are flawed. From Deci and Ryan's work on how incentives fail to Lave and Wenger's studies of communities of practice, learning appears to be a highly situated activity that resists supposedly rational procedural schemes. As Yochai Benkler points out in his multidisciplinary work on cooperation and self-interest, *The Penguin and the Leviathan*, important factors that guide human decision making reflect values that include autonomy, fairness, altruism, empathy, solidarity, and loyalty.[48] Interpretive framing, trust, reputation within a network, access to face-to-face and multichannel communication, and what Benkler calls "susceptibility to symbolic performances" play critical roles. His criticism of the oversimplification inherent in theories of behaviorism and scientific management can be applied directly to the questionable claims of the learning sciences that are similarly grounded in faulty assumptions about reward systems and efficiency. Yet much as we might know from empirical science that dieting does little to foster healthy behavior, while the weight loss industry reaps profits from naïve consumers, the learning sciences thrive because they deploy commonsensical sounding truisms. Recent discoveries in cognitive science emphasize the importance of emotions and embodied experiences in education, yet the learning sciences insist that impersonal algorithms and gadgets should rule.

As a practitioner of the learning arts, I have not bothered to debate the learning sciences point-by-point in this volume. I think that many of their basic premises are flawed. I believe that the rhetoric around the learning sciences is disingenuous. When I watch their experiments, I do not see success. Fortunately, I do see success in many sectors of the digital university, where students are developing, learning to make decisions, testing out identities, and seeking truth. I also see success where research on learning and on the use of computational media—sometimes separately—is flourishing in the academy. I hope that this book contributes to these efforts and to the public's understanding and support.

Notes

Introduction

1. Susan Matt and Luke Fernandez, "Before MOOCs, 'Colleges of the Air,'" *Chronicle of Higher Education: The Conversation*, April 23, 2013, http://chronicle.com/blogs/conversation/2013/04/23/before-moocs-colleges-of-the-air/.

2. Adrian Mackenzie, *Wirelessness Radical Empiricism in Network Cultures* (Cambridge, MA: MIT Press, 2010), presents the ideology of wirelessness as identified with the radical empiricism of William James.

3. Nicole Starosielski, "'Warning: Do Not Dig': Negotiating the Visibility of Critical Infrastructures," *Journal of Visual Culture* 11, no. 1 (April 1, 2012): 38–57.

4. Leopoldina Fortunati, "The Mobile Phone: Towards New Categories and Social Relations," *Information, Communication & Society* 5, no. 4 (2002): 513–528.

5. Lisa Nakamura, "Economies of Digital Production in East Asia," *Media Fields Journal* 5 (2012), http://www.mediafieldsjournal.org/economies-of-digital/.

6. Nicholas Negroponte, *Being Digital* (New York: Knopf, 1995).

7. Matthew Kirschenbaum, *Mechanisms: New Media and the Forensic Imagination* (Cambridge, MA: MIT Press, 2008).

8. Paul Dourish and Genevieve Bell, *Divining a Digital Future: Mess and Mythology in Ubiquitous Computing* (Cambridge, MA: MIT Press, 2011), 4–5.

9. Ellen Lupton and J. Abbott Miller, *The Bathroom, the Kitchen, and the Aesthetics of Waste: A Process of Elimination* (Princeton: Princeton University Press, 1996).

10. See, e.g., Laurent Mannoni and Richard Crangle, *The Great Art of Light and Shadow: Archeology of the Cinema* (Exeter, Devon: University of Exeter Press, 2000), the Peabody Magic Lantern Collection as San Diego State University, and the work of Amy-Claire Huestis.

11. LaMond F. Beatty, *Filmstrips* (Englewood Cliffs, NJ: Educational Technology Publications, 1981).

12. Jerry D. Sparks, *Overhead Projection* (Englewood Cliffs, NJ: Educational Technology Publications, 1981).

13. John Law, *After Method: Mess in Social Science Research* (New York: Routledge, 2004).

14. John Law, "Notes on the Theory of the Actor-Network: Ordering, Strategy, and Heterogeneity," *Systems Practice* 5, no. 4 (August 1, 1992): 382.

15. Claude Elwood Shannon, *The Mathematical Theory of Communication* (Urbana: University of Illinois Press, 1949).

16. J. C. R. Licklider and Robert W. Taylor, "The Computer as a Communication Device," *Science and Technology* 76 (1968): 21–31.

17. David Golumbia, *The Cultural Logic of Computation* (Cambridge, MA: Harvard University Press, 2009), 11. It is noteworthy that Golumbia cites Theodor Roszak's classic book about the computerizing of learning, *The Cult of Information: A Neo-Luddite Treatise on High Tech, Artificial Intelligence, and the True Art of Thinking* (Berkeley: University of California Press, 1994) in his critique of "the project of instrumental reason."

18. Paul A. David, "Clio and the Economics of QWERTY," *American Economic Review* 75, no. 2 (May 1985): 332–337.

19. S. Eugene Michaels, "Qwerty Versus Alphabetic Keyboards as a Function of Typing Skill," *Human Factors: The Journal of the Human Factors and Ergonomics Society* 13, no. 5 (October 1, 1971): 419–426.

20. "NSF Future Internet Architecture Project."

21. S. J. Liebowitz and Stephen E. Margolis, "Path Dependence, Lock-in, and History," *Journal of Law, Economics, and Organization* 11, no. 1 (April 1995): 205–226.

22. For more about the history of digital rhetoric, see "Hacking Aristotle: What is Digital Rhetoric," in Elizabeth Losh, *Virtualpolitik: An Electronic History of Government Media-making in a Time of War, Scandal, Disaster, Miscommunication, and Mistakes* (Cambridge, MA: MIT Press, 2009) and Douglas Eyman's forthcoming book on the subject.

23. "Join the Blog Carnival: 'What Does Digital Rhetoric Mean to Me?,'" n.d., http://www.digitalrhetoriccollaborative.org/2012/05/17/join-the-blog-carnival-what-does-digital-rhetoric-mean-to-me/.

24. For example, in 2007 I moderated and helped organize a public panel at my university called "Serious Play: The Practices of Everyday Life in Video Games and Virtual Worlds," which was largely devoted to what I thought were under-discussed

subjects in the study of new media: routine, repetition, chores, contractual obligations, boredom, and waiting. Many of the speakers discussed how university life was similarly full of required tasks that were equally mundane, even when academic inquiry was presented as an exciting adventure. Strangely, a reporter from a local radio station called to find out more about my panel on "violence and video games" and made sex and violence the focus of the story that subsequently aired.

25. Ian Bogost, *Persuasive Games: The Expressive Power of Videogames* (Cambridge, MA: MIT Press, 2007).

26. Larry Cuban, *Teachers and Machines: The Classroom Use of Technology Since 1920* (New York: Teachers College Press, 1986).

1 What They Learn in College

1. Elizabeth Losh, "Cheaters Never Prosper," *Virtualpolitik* (blog), July 9, 2008, http://virtualpolitik.blogspot.com/2008/07/cheaters-never-prosper.html.

2. "How to Cheat on Any Test," YouTube video, 3:24, posted by "Household-Hacker," March 17, 2008, http://www.youtube.com/watch?v=91lQK5SCzlQ.

3. "How to Cheat on a Test," YouTube video, 3:46, posted by "swtxwishes," September 26, 2007, http://www.youtube.com/watch?v=rCzbq3XCM9o (video no longer publicly available).

4. "How to Cheat on a Test," YouTube video, 1:42, posted by "dbbvlog," February 3, 2008, http://www.youtube.com/watch?v=tGhOYbPgETQ.

5. "カンニング実践講座2 (How to Cheat on an Exam 2)," YouTube video, 3:55, posted by "StronBluebookProject," April 17, 2007, http://www.youtube.com/watch?v=eWM5Rqjlx0I.

6. Andrea L. Foster, "Students Show How to Cheat via YouTube," *Wired Campus* (blog), July 11, 2008, http://chronicle.com/wiredcampus/article/3160/students-show-how-to-cheat-via-youtube.

7. Icess Fernandez, "Videos Teach Students Deception," *Shreveport Times*, September 17, 2008, http://www.shreveporttimes.com.

8. Icess Fernandez, "Kids Spread Cheating Methods on YouTube," *Chicago Sun-Times*, October 1, 2008.

9. "To Catch a Predator," which aired on *NBC Dateline* from 2004 to 2007, probably epitomized this panic. See "Potential Predators Go South in Kentucky," n.d., http://www.msnbc.msn.com/id/10912603/. The PBS show *Frontline* devoted a segment that was critical of the panic over "stranger danger" featuring danah boyd. See "Predator Panic," February 2, 2010, http://www.pbs.org/wgbh/pages/frontline/digitalnation/relationships/predators-bullies/predator-panic.html.

10. The 2010 suicide of Rutgers student Tyler Clementi—seemingly in connection with online humiliation that included roommate Dharun Ravi recording a homosexual liaison between Clementi and another man on a webcam and broadcasting their sexual activity on Twitter—caused many to focus on online bullying as an issue, although Wikipedia has an entry on "cyberbullying" that dates back to 2005. For a protest against the undue influence of this panic, see Nick Gillespie, "Stop Panicking about Bullies," *Wall Street Journal*, April 2, 2012, http://online.wsj.com/article/SB10001424052702303404704577311664105746848.html.

11. "High-Tech Cheating," *Good Morning America*, 5:29, October 10, 2008, http://abcnews.go.com/video/playerIndex?id=6002101.

12. Mizuko Ito, *Hanging Out, Messing Around, and Geeking Out: Kids Living and Learning with New Media* (Cambridge, MA: MIT Press, 2009).

13. Ibid., 221.

14. Mia Consalvo, *Cheating: Gaining Advantage in Videogames* (Cambridge, MA: MIT Press, 2007).

15. James Paul Gee, *What Video Games Have to Teach Us about Learning and Literacy*, rev. and updated ed. (New York: Palgrave, 2007).

16. Ian Bogost, *Persuasive Games: The Expressive Power of Videogames* (Cambridge, MA: MIT Press, 2007).

17. Betsy DiSalvo and Amy Bruckman, "Race and Gender in Play Practices: Young African American Males," *FDG '10: Proceedings of the Fifth International Conference on the Foundations of Digital Games*, http://portal.acm.org/citation.cfm?id=1822356.

18. Sneha Veeragoudar Harrell and D. Fox Harrell, "Exploring the Potential of Computational Self-Representations for Enabling Learning: Examining At-Risk Youths' Development of Mathematical/Computational Agency," UC Irvine: Digital Arts and Culture 2009, http://escholarship.org/uc/item/4b6913rb.

19. "Tracy Fullerton on Pathfinder Project," YouTube video, 2:48, posted by "hreingold," November 18, 2009, http://www.youtube.com/watch?v=_Em-z7N_kgo.

20. For a more complete discussion of virtual worlds as refuges, see Edward Castronova, *Exodus to the Virtual World: How Online Fun Is Changing Reality* (New York: Palgrave, 2008). As an economist doing experiments in the Arden project, which will be discussed later in this book, Castronova has also claimed that real-world financial literacy is not completely disconnected from virtual forms of consumption.

21. James Paul Gee, *The Anti-Education Era: Creating Smarter Students through Digital Learning* (New York: Palgrave, 2013), 114.

22. Ibid., 115.

23. Foster, "Students Show How to Cheat via YouTube."

24. Margaret Spellings, *The Secretary of Education's Commission on the Future of Higher Education,* http://www2.ed.gov/about/bdscomm/list/hiedfuture/reports/final-report.pdf.

25. "Science Research Associates" n.d., http://en.wikipedia.org/wiki/Science_Research_Associates.

26. "National Survey of Student Engagement," n.d., http://nsse.iub.edu/.

27. "CCSSE—Community College Survey of Student Engagement," n.d., http://www.ccsse.org/.

28. "HSSSE Intro," n.d., http://ceep.indiana.edu/hssse/.

29. See, e.g., Clifford Geertz, *Local Knowledge: Further Essays in Interpretive Anthropology* (New York: Basic Books, 1983) and Pierre Bourdieu, *Outline of a Theory of Practice* (Cambridge: Cambridge University Press, 1977).

30. The antonym term "digital immigrant" is also weighted with the ideological baggage of national politics and the particular cultural imaginaries of a nation-state.

31. For more on this subject, see Jonathan Alexander and Elizabeth Losh, "Whose Literacy Is It Anyway? Examining a First-Year Approach to Gaming Across Curricula," *Currents in Electronic Literacy* 2010, http://currents.cwrl.utexas.edu/2010/alexander_losh_whose-literacy-is-it-anyway.

32. See http://processing.org for more.

33. For more about the virtues of Processing, see Michael Mateas, "Procedural Literacy: Educating the New Media Practitioner," *On the Horizon* 13, no. 2 (2005): 101–111.

34. John Palfrey and Urs Gasser, *Born Digital: Understanding the First Generation of Digital Natives* (New York: Basic Books, 2010), 1, 2, 4, and 6.

35. S. Craig Watkins, *The Young and the Digital: What the Migration to Social-Network Sites, Games, and Anytime, Anywhere Media Means for Our Future* (Boston: Beacon Press, 2009).

36. A search of the online *New York Times* using the term "digital natives" brings up over a hundred and fifty articles, although many of them question the accuracy of the term, such as Sarah Perez, "So-Called 'Digital Natives' Not Media Savvy, New Study Shows," *New York Times*, July 29, 2010, http://www.nytimes.com/external/readwriteweb/2010/07/29/29readwriteweb-so-called-digital-natives-not-media-savvy-n-74704.html.

37. Eszter Hargittai, "Digital Na(t)ives? Variation in Internet Skills and Uses among Members of the Net Generation," *Sociological Inquiry* 80, no. 1 (2010): 92–113.

38. "The Playboy Interview: Marshall McLuhan," *Playboy*, March 1969, http://www.digitallantern.net/mcluhan/mcluhanplayboy.htm.

39. Television Information Office (U.S.) Library, *Television and Education: A Bibliography* (New York: The Library, 1960).

40. Siva Vaidhyanathan, "Generational Myth," *Chronicle of Higher Education*, September 19, 2008, http://chronicle.com/article/Generational-Myth/32491.

41. "Benefits of Clicker Technology," http://www.engaging-technologies.com/clicker-technology.html.

42. Matt Richtel, "Growing Up Digital, Wired for Distraction," *New York Times*, November 21, 2010, http://www.nytimes.com/2010/11/21/technology/21brain.html.

43. Cathy N. Davidson, *Now You See It: How the Brain Science of Attention Will Transform the Way We Live, Work, and Learn* (New York: Viking, 2011), 17.

44. "Teacher Breaks Student's Phone," YouTube video, 0:36, posted by "pvtvfilm," June 8, 2010, http://www.youtube.com/watch?v=TEbPLWB3iZE.

45. "'I Have a Strict Texting Policy' (Professor Destroys Cell Phone)," YouTube video, 2:21, posted by "paulatravisphillips," August 28, 2010, http://www.youtube.com/watch?v=9UtRsGU6pVs.

46. Lauren Dielman, "Texting in Class Seen as Disruptive to Students, Teachers," *Northern Star Online*, June 22, 2012, http://northernstar.info/campus/news/article_1f16a0d4-4550-11e1-92b4-0019bb30f31a.html.

47. Kieran Mullen's Homepage, http://www.nhn.ou.edu/~kieran.

48. "Why Students Don't Text Message in My Class," YouTube video, 2:02, posted by "SeattlePacific," December 22, 2010, http://www.youtube.com/watch?v=NZ5Zahh9akI.

49. "Cornell Professor Outbursts at a Student's 'Overly Loud' Yawn," YouTube video, 2:29, posted by "opieboyboy," November 5, 2010, http://www.youtube.com/watch?v=QuLaQoQP9oo.

50. Amy Johnson and Trevor Smith, "Professor Arrested for Battery," *Spectator*, March 31, 2011, http://www.vsuspectator.com/2011/03/31/professor-arrested-for-battery/.

51. Scott Jaschik, "News: Laptop and Battery (Arrest)," *Inside Higher Ed*, April 4, 2011, http://www.insidehighered.com/news/2011/04/04/professor_arrested_for_battery_after_he_shut_laptop_of_a_student_in_class.

52. "Team Rybicki (3)," n.d., http://www.facebook.com/groups/teamrybicki/.

53. Trevor Smith, "VSU Professor Found Not Guilty," *Spectator*, August 26, 2011, http://www.vsuspectator.com/2011/08/26/vsu-professor-found-not-guilty/.

54. Scott Jaschik, "News: Not Guilty ... and Not Long Employed," *Inside Higher Ed*, August 25, 2011, http://www.insidehighered.com/news/2011/08/25/jury_rejects_charges_against_professor_in_case_of_student_laptop.

55. Ibid.

56. "New Wrinkle in 'No Cell Phones in Class,'" *Courthouse News Service*, September 29, 2010, http://www.courthousenews.com/2010/09/29/30652.htm.

57. "FAU Student Threatens to Kill Professor and Classmates," 2013, *University Press*, http://upressonline.com/2012/03/fau-student-threatens-to-kill-professor-and-classmates/.

58. "Black Woman Meltdown at Florida Atlantic Interviewed about Trayvon Martin Shortly before Episode and the Full Rant Video [Video]," 2013, *Bossip*, http://bossip.com/560571/black-woman-who-had-meltdown-at-florida-atlantic-interviewed-about-trayvon-martin-shortly-before-episode-and-the-full-rant-video-video69691/.

59. "RE: Girl at FAU Goes Crazy and Gets Tazed by Cops," March 21, 2012, http://www.youtube.com/watch?v=GHr5L4y6-g8.

60. "Don't Mess with This Teacher!," September 23, 2010, http://www.youtube.com/watch?v=q9cbM18bTj4.

61. "Prof Smashes Laptop: What REALLY Happened," April 7, 2008, http://www.youtube.com/watch?v=spKuQAdf5r8.

2 The War on Learning

1. Nicholas A. Christakis and James H. Fowler, *Connected: The Surprising Power of Our Social Networks and How They Shape Our Lives* (New York: Little, Brown, 2009).

2. danah boyd, "Spectacle at Web2.0 Expo ... from My Perspective" http://www.zephoria.org/thoughts/archives/2009/11/24/spectacle_at_we.html.

3. Ibid.

4. In winter 2012 I taught the class again with more emphasis on student agency in digital content-creation. It received excellent course evaluations and became a model of upper-division writing instruction on the campus.

5. For more on this subject, see Howard Rheingold, *Smart Mobs: The Next Social Revolution* (Cambridge, MA: Perseus, 2003) and Jodi Dean, Jon W. Anderson, and Geert Lovink, *Reformatting Politics: Information Technology and Global Civil Society* (New York: Routledge, 2006).

6. JonnyHu, "Internet Dies in Egypt," *CAT 125*, January 28, 2011, http://excelsior.ucsd.edu/siorgs/cat125/?p=2694.

7. sdrake, "Hey! He Gypt My Facebook!," *CAT 125*, February 3, 2011, http://excelsior.ucsd.edu/siorgs/cat125/?p=3303.

8. For more on human rights remixes about the Arab Spring, see Sam Gregory and Elizabeth Losh, "Remixing Human Rights: Rethinking Civic Expression, Safety, Privacy and Consent in Online and Mobile Video," *First Monday* 17, no. 8 (August 2012), http://www.firstmonday.org/ojs/index.php/fm/article/view/4104/3279.

9. "Egypt, *Twitter & What* This Means for Us," 2011, http://www.youtube.com/watch?v=iUK--TDCOpo&feature=youtube_gdata_player.

10. Ibid.

11. Brian Goldfarb and Alexandra Juhasz, "Distraction Span: Introduction," *New Everyday* (2011), http://mediacommons.futureofthebook.org/tne/pieces/distraction-span-introduction.

12. Ibid.

13. See Elizabeth Losh, "Silent Readings: Lessons in Objectivist Poetics for Contemporary American Poetry," 1998 (dissertation), which describes "ten years of heated debate about the 'death of poetry,' a phenomenon supposedly caused by professional writing programs and the universities that sponsored them." It explains how Joseph Epstein's much-discussed essay "Who Killed Poetry?" warns that "the entire enterprise of poetic creation seems threatened by having been taken out of the world, chilled in the classroom, and vastly overproduced by men and women who are licensed to write it by degree if not necessarily by talent or spirit." The responses to Epstein ranged from assertions of poetry's continued good health (Donald Hall's *Death to the Death of Poetry*) to an acceptance of poetry's demise that posits an afterlife for it sustained by a small canon of contemporary poets of great merit (Vernon Shetley's *After the Death of Poetry*) to orders for stern remedies of technical discipline and financial independence from the university (Dana Gioia's *Can Poetry Matter?*).

14. National Endowment for the Arts, *Reading at Risk: A Survey of Literary Reading in America: Executive Summary* (Washington, DC: National Endowment for the Arts, 2004), vii–xiii.

15. For more on so-called "secondary orality," see Walter Ong, *Orality and Literacy: The Technologizing of the Word* (New York: Methuen, 1982).

16. Sharon Begley, "Culture: The Dumbest Generation? Don't Be Dumb," *Newsweek*, May 24, 2008, http://www.thedailybeast.com/newsweek/2008/05/24/the-dumbest-generation-don-t-be-dumb.html.

17. *Reading at Risk* notes that females are more likely to read literature than males. For the idealization of female readers and the imagery of contemplation in the

hortus conclusus that appears in Sven Birkerts's nostalgia for a predigital age, see Elizabeth Losh, "Between the Angel and the Book: The Female Reading Subject of Early Modern Flemish Annunciation Painting," 2001, https://eee.uci.edu/faculty/losh/pubs/Angel.htm.

18. Mark Bauerlein, *The Dumbest Generation: How the Digital Age Stupefies Young Americans and Jeopardizes Our Future* (New York: Tarcher, 2008).

19. To be fair to Bauerlein, he recently edited a collection that featured critics from the opposing camp of digital learning enthusiasts. See Mark Bauerlein, *The Digital Divide: Arguments for and Against Facebook, Google, Texting, and the Age of Social Networking* (New York: Tarcher, 2011).

20. See, e.g., Sonia Livingstone, *Children and the Internet: Great Expectations, Challenging Realities* (Cambridge: Polity, 2009).

21. Bauerlein, *The Dumbest Generation*, 133.

22. When I visit discussion-oriented classes to evaluate the instructor, I keep count in my notes of my observation not only of how many students speak to the instructor and how often, but also how students address each other in the space. Teachers who succeed have students talking to each other about course material; teachers who fail tend to be those who can only manage one channel of communication at a time.

23. Mizuko Ito, *Hanging Out, Messing Around, and Geeking Out: Kids Living and Learning with New Media* (Cambridge, MA: MIT Press, 2010).

24. See Chris Anderson, *The Long Tail: Why the Future of Business Is Selling Less of More*, rev. and updated ed. (New York: Hyperion, 2008).

25. Paul Graham, *Hackers and Painters: Big Ideas from the Computer Age* (Farnham: O'Reilly, 2004).

26. danah boyd, *Why Youth (Heart) Social Network Sites: The Role of Networked Publics in Teenage Social Life* (The MacArthur Foundation Digital Media and Learning Initiative, 2008).

27. Manuel Castells, *The Power of Identity* (Malden, MA: Blackwell, 1997).

28. Nicholas Carr, *The Shallows: What the Internet Is Doing to Our Brains* (New York: W. W. Norton, 2010), 7.

29. See the work of Stella de Bode, e.g., "Language after Hemispherectomy," *Brain Cognition* 43, no. 1–3 (2000): 135–138.

30. N. Katherine Hayles, *How We Think: Transforming Power and Digital Technologies* (Chicago: University of Chicago Press, 2011), 57. The author wishes to thank Professor Hayles for giving her access to manuscript copy.

31. L. Vygotskiĭ, *Mind in Society: The Development of Higher Psychological Processes* (Cambridge, MA: Harvard University Press, 1978).

32. Deborah H. Holdstein, "'Writing Across the Curriculum' and the Paradoxes of Institutional Initiatives," *Pedagogy* 1, no. 1 (2001): 37–52.

33. Frank Donoghue, *The Last Professors: The Corporate University and the Fate of the Humanities*, 3rd ed. (New York: Fordham University Press, 2008), 108.

34. Toby Miller, *Blow Up the Humanities* (Philadelphia: Temple University Press, 2012), 1–2.

35. Ibid., 105.

36. Derek Curtis Bok, *Our Underachieving Colleges: A Candid Look at How Much Students Learn and Why They Should Be Learning More* (Princeton, NJ: Princeton University Press, 2006).

37. Mark C. Taylor, "End the University as We Know It," *New York Times*, April 27, 2009, Opinion sec., http://www.nytimes.com/2009/04/27/opinion/27taylor.html.

38. My closest colleagues found themselves unable to figure out how to situate their own research in Taylor's supposedly more inclusive schema. For example, my own work could be placed in "Media," "Information," and "Networks," which is hardly an improvement in number on the three departments in which I currently teach: Communication, Literature, and Visual Arts.

39. Many professional journalists, such as Pulitzer Prize winner Jeff Brazil, who visited my digital journalism class in spring 2012, argue that consolidation of ownership is to blame for the decline of the field, as publicly traded multinational corporations replace family-run enterprises. Brazil also argues that individual bloggers can't compete with investigative journalists who once had the backing of broadcast networks or newspaper chains, because bloggers don't have the legal resources to file lawsuits to gain access to key documents or to fight frivolous libel or defamation claims by the powerful.

40. Cathy Davidson and David Theo Goldberg, *The Future of Learning Institutions in a Digital Age* (Cambridge, MA: MIT Press, 2009), 5.

41. Ibid., 14.

42. Ibid., 16.

43. Tom Scheinfeldt and Dan Cohen, "Hacking the Academy," n.d., http://hackingtheacademy.org/what-this-is-and-how-to-contribute/.

44. Ibid.

45. Trebor Scholz, "Learning through Digital Media," 2011, http://learningthroughdigitalmedia.net/.

46. Florian Cramer, "Is There Hope?" presented at the Mobility Shifts, New York, October 15, 2011.

47. Bien, Waldo. "Statement for and on Behalf of FIUWAC," n.d., http://www.fiuwac.com/html/fiuwac_statement.html.

48. "FREE U LOG," n.d., http://www.copenhagenfreeuniversity.dk/log.html.

49. Gary Hall, "The Free, 'Libre,' University," *Vimeo*, February 24, 2010, http://vimeo.com/9711800.

50. "What Is the ROU?" *Really Open University*, n.d., http://reallyopenuniversity.wordpress.com/what-is-the-rou/.

51. "About the University," n.d., http://universityofthepoor.org/?page_id=7.

52. "About Edufactory," n.d., http://www.edu-factory.org/wp/about/.

53. Jacques Derrida, "The Future of the Profession or the University Without Condition (Thanks to the 'Humanities,' What Could Take Place Tomorrow)," in *Jacques Derrida and the Humanities: A Critical Reader*, ed. Tom Cohen (Cambridge: Cambridge University Press, 2001).

54. Hall, "The Free, 'Libre,' University." .

55. Michael Hardt and Antonio Negri, *Multitude: War and Democracy in the Age of Empire* (New York: Penguin, 2004).

56. See http://thepublicschool.org/.

57. Douglas Thomas and John Seely Brown, *A New Culture of Learning: Cultivating the Imagination for a World of Constant Change* (Lexington, KY: CreateSpace, 2011), 18.

58. Ibid., 92.

59. Ibid., 22.

60. Interview with the author.

61. Thomas and Brown, *A New Culture of Learning*, 59.

62. See Anke Schwittay and Paul Braund of the Institute for Money, Technology, and Financial Inclusion, "'Democratizing Capital': Digital Lending Networks, Mobile Technologies and Women's Solidarity Groups in Chiapas, Mexico and Guatemala," described at http://virtualpolitik.blogspot.com/2010/10/placing-trust-in-kiva.html.

63. Thomas and Brown, *A New Culture of Learning*, 45.

64. Geert Lovink, *Zero Comments: Blogging and Critical Internet Culture* (New York: Routledge, 2008).

65. Jeffrey J. Selingo, *College (Un)bound: The Future of Higher Education and What It Means for Students* (Boston: Houghton Mifflin, 2013).

66. Chuck Rybak, "On Jeffrey Selingo's College (Un)bound," *Sad Iron*, June 26, 2013, http://www.sadiron.com/on-jeffrey-selingos-college-unbound/.

67. Anya Kamenetz, *DIY U: Edupunks, Edupreneurs, and the Coming Transformation of Higher Education* (White River Junction, VT: Chelsea Green, 2010).

68. Ibid., ix.

69. Ibid., 34.

70. Ibid., 52.

71. Ibid., 55.

72. Ibid., vii.

73. Ibid., 103.

74. Ibid., 92.

75. "College, Inc.," n.d., http://www.pbs.org/wgbh/pages/frontline/collegeinc/view/.

76. Jeff Jarvis, "TEDxNYed: This Is Bullshit—BuzzMachine," *BuzzMachine*, March 8, 2010, http://buzzmachine.com/2010/03/08/tedxnyed-this-is-bullshit/.

77. Ibid.

78. "TEDxNYED—Jeff Jarvis—03/06/10," 2010, http://www.youtube.com/watch?v=rTOLkm5hNNU.

79. Kathleen Blake Yancey, "Postmodernism, Palimpsest, and Portfolios: Theoretical Issues in the Representation of Student Work," *College Composition and Communication* 55, no. 4 (June 2004): 738–761.

80. Jeff Jarvis, *What Would Google Do?* (New York: Collins Business, 2009), 222.

81. Ibid., 214.

82. Siva Vaidhyanathan, *The Googlization of Everything (and Why We Should Worry)* (Berkeley: University of California Press, 2011).

83. Ray Oldenburg, *The Great Good Place: Cafés, Coffee Shops, Community Centers, Beauty Parlors, General Stores, Bars, Hangouts, and How They Get You through the Day* (New York: Paragon House, 1989).

84. "The Classroom as a Sacred Space—Siva Vaidhyanathan," 2010, http://fora.tv/2010/04/21/Siva_Vaidhyanathan_The_Classroom_Is_Sacred#fullprogram.

85. Ibid.

86. "How Should the University Evolve? A Conversation about the Future of Higher Education," Baruch College, 2010, http://vimeo.com/17140344.

87. "NCAT Homepage," n.d., http://www.thencat.org/.

88. The question specifically addressed the value of "The Great Courses—Audio and Video Lectures from the World's Best Professors," n.d., http://www.thegreatcourses.com/greatcourses.aspx.

89. The reference point is, of course, Habermas's classic work about bourgeois society in coffeehouses, contained in *The Structural Transformation of the Public Sphere* (Cambridge, MA: MIT Press, 1989).

90. Kamenetz, *DIY U*, 57.

91. Eric S. Raymond, *The Cathedral and the Bazaar: Musings on Linux and Open Source by an Accidental Revolutionary* (Sebastopol, CA: O'Reilly, 2001).

92. "The Education Bazaar," n.d., http://teachers4schools.com/open/about/.

93. J0chen24, "The Cathedral and the Bazaar," *Cardinal Lawyer*, November 1, 2007, http://www.law.louisville.edu/cardinallawyer/node/25.

94. Jarvis, *What Would Google Do?*, 210.

95. Michael Wesch, "A Vision of Students Today," 2007, http://www.youtube.com/watch?v=dGCJ46vyR9o.

96. Mark Marino, "A (Re)Vision of Students Today: Remixing Wesch at WRT: Writer Response Theory," *WRT: Writer Response Theory*, January 30, 2008, http://writerresponsetheory.org/wordpress/2008/01/20/a-revision-of-students-today-remixing-wesch/.

97. "(Re)Visions of Students Today," 2008, http://www.youtube.com/watch?v=Ln6WUy29fAA.

98. Michael Wesch, "(Re)Visions of Students Today," *Digital Ethnography*, January 21, 2008, http://mediatedcultures.net/ksudigg/?p=133.

99. Kevin Yamamoto, "Banning Laptops in the Classroom: Is It Worth the Hassles?" *SSRN eLibrary* (n.d.), http://papers.ssrn.com/sol3/papers.cfm?abstract_id=1078740.

100. Carrie B. Fried, "In-class Laptop Use and Its Effects on Student Learning," *Computers & Education* 50, no. 3 (April 2008): 906–914.

101. See John Schwartz, "Professors Vie with Web for Class's Attention," *New York Times*, January 2, 2003, http://www.nytimes.com/2003/01/02/us/professors-vie-with-web-for-class-s-attention.html.

102. Daniel de Vise, "College Inc.—Laptops in Class: Tool or Distraction?" *Washington Post*, March 9, 2010, http://voices.washingtonpost.com/college-inc/2010/03/laptops_in_class_have_we_creat.html.

103. Netop, "How to Restrict Internet Access and Guide Students Online," http://www.netop.com/classroom-management-software/products/netop-vision/restrict-internet-access.htm.

104. For more about the concept of "holding power" in computational media, see Sherry Turkle, *Video Games and Computer Holding Power* (New York: Simon & Schuster, 1984).

105. Such intrusive Web control could violate the concept of "negative politeness" that is developed in Penelope Brown and Stephen C. Levinson, *Politeness: Some Universals in Language Usage* (Cambridge: Cambridge University Press, 1987), in which politeness is constituted by not interfering in the activities, goals, or private space of others.

106. Joseph Janangelo, "Technopower and Technoppression: Some Abuses of Power and Control in Computer-assisted Writing Environments," *Computers and Composition* 9, no. 1 (November 1991): 47–64.

107. "Children's Internet Protection Act," FCC, n.d., http://www.fcc.gov/guides/childrens-internet-protection-act.

108. Henry Jenkins and Elizabeth Losh, "Can Public Education Coexist with Participatory Culture?," *Knowledge Quest*, Sept.–Oct. 2012.

109. David A. Hoekema, "Beyond *in Loco Parentis*? Parietal Rules and Moral Maturity," in Steven M. Cahn, *Morality, Responsibility, and the University: Studies in Academic Ethics* (Philadelphia: Temple University Press, 1990). See also David A. Hoekema, *Campus Rules and Moral Community: In Place of In Loco Parentis* (Lanham, MD: Rowman & Littlefield, 1994).

110. Ivan Tribble, "Bloggers Need Not Apply," *Chronicle of Higher Education*, July 8, 2005, http://chronicle.com/article/Bloggers-Need-Not-Apply/45022/.

111. Elizabeth Losh, "Hacktivism and the Humanities: Programming Protest in the Era of the Digital University," in *Debates in the Digital Humanities*, ed. Matthew K. Gold (Minneapolis: Minnesota University Press, 2012), 161–186.

112. "The English TA Experience [1 of 2]," 2008, http://www.youtube.com/watch?v=kVaW3rZWK5E.

113. "The English TA Experience by Iowa State Students," http://www.new.facebook.com/group.php?gid=35107035121.

114. "College Sued Over 'Drunken Pirate' Sanctions," April 26, 2007, http://www.thesmokinggun.com/archive/years/2007/0426072pirate1.html.

115. "Valdosta State University: Student Expelled for Peacefully Protesting Parking Garages," n.d., http://www.thefire.org/index.php/case/751.html.

116. Andy Guess, "Maybe He Shouldn't Have Spoken His Mind," *Inside Higher Ed*, January 11, 2008, http://www.insidehighered.com/news/2008/01/11/valdosta.

117. Jill Laster, "Student Punished for Facebook Group Starts $10-Million Lawsuit," Wired Campus, March 22, 2010, http://chronicle.com/blogs/wiredcampus/student-punished-for-facebook-group-starts-10-million-lawsuit/21974.

118. Louise Brow, "Student Faces Facebook Consequences," Toronto Star, March 6, 2008, http://www.thestar.com/News/GTA/article/309855. See Clay Shirky, *Cognitive Surplus: How Technology Makes Consumers into Collaborators* (New York: Penguin, 2011), for more on the Ryerson case.

119. Oklahoma State University, "Appropriate Computer Use," http://it.okstate.edu/policies/pol_app.php/.

120. See Arthur W. Chickering, Zelda F. Gamson, and Susan J. Poulse, *Seven Principles for Good Practice in Undergraduate Education* (Racine, WI: Johnson Foundation, 1987) and Arthur W. Chickering and Zelda F. Gamson, *Applying the Seven Principles for Good Practice in Undergraduate Education* (San Francisco, CA: Jossey-Bass, 1991).

121. "Project Talent," n.d., http://www.projecttalent.org/.

122. A. W. Chickering and S. C. Ehrmann, "Implementing the Seven Principles: Technology as Lever," *AAHE BULLETIN* 49, no. 2 (1996): 3–6.

123. Ibid.

124. See Alexandre Enkerli, "Brewing Cultures: Craft Beer and Cultural Identity in North America," presented at the joint conference of the Association for the Study of Food and Society (ASFS) and the Agriculture, Food, and Human Values Society (AFHVS), Boston, June 2006.

125. Jane Jacobs, *The Death and Life of Great American Cities* (New York: Modern Library, 2011).

126. Carol A. Twigg, *Improving Learning and Reducing Costs: Redesigning Large Enrollment Courses*, http://www.thencat.org/Monographs/ImpLearn.html.

127. Paulo Freire, *Pedagogy of the Oppressed*, new rev. 20th anniv. ed. (New York: Continuum, 1993).

3 On Camera: The Baked Professor Makes His Debut

1. "Baked Professor Is Burned on Blog, Canned by College," *Chronicle of Higher Education*, September 28, 2006, http://chronicle.com/news/article/1057/baked-professor-is-burned-on-blog-canned-by-college.

2. Jack Stripling, "Erratic Behavior Causes UF to Put Lecturer on Leave," *Gainesville Sun*, September 30, 2006, http://gainesville.com/apps/pbcs.dll/article?AID=/20060930/LOCAL/60930001/1078/news.

3. Eric S. Raymond, *The Cathedral and the Bazaar: Musings on Linux and Open Source by an Accidental Revolutionary* (Sebastopol, CA: O'Reilly, 2001).

4. For more about Max Weber's classic work on bureaucracy, see Charles Camic, Philip S. Gorski, and David M. Trubek, *Max Weber's Economy and Society: A Critical Companion* (Stanford, CA: Stanford University Press, 2005). Weber also notes that modern bureaucracies maintain permanent files, separate personal property from office property, and promote the idea that employment functions in the arc of a career.

5. "Baked Biz School Prof, Part 2," 2006 (video no longer available online).

6. Ibid.

7. David F. Noble, *Digital Diploma Mills: The Automation of Higher Education* (New York: Monthly Review Press, 2001).

8. Frank Donoghue, *The Last Professors: The Corporate University and the Fate of the Humanities* (New York: Fordham University Press, 2008).

9. Eric Breitbart and Taylor Project, *Clockwork Documentary Film*, 2011. Of course, making digital copies of the film could be defended by the principle of fair use in educational contexts, just as academic freedom might justify the criticism of the educational system that follows the screening.

10. Taylor's original work is remarkably rhetorically complex, beginning with its introduction praising the contemporary environmental movement and the preservation efforts of then-president Theodore Roosevelt. See Frederick Winslow Taylor, *The Principles of Scientific Management* (New York: Norton, 1967).

11. Noble, *Digital Diploma Mills*, 33.

12. Donoghue, *The Last Professors*, 7–8.

13. See, e.g., Martha Lampland and Susan Leigh Star, *Standards and Their Stories: How Quantifying, Classifying, and Formalizing Practices Shape Everyday Life* (Ithaca: Cornell University Press, 2009).

14. Jürgen Habermas and Thomas MacCarthy, *The Theory of Communicative Action*, vol. 2: *Lifeworld and System: A Critique of Functionalist Reason* (Cambridge: Polity Press, 1995).

15. See Alexandra Juhasz on popularity in "TOUR #3: POPULARITY! Who Doesn't Want to be Prom Queen?," 2008, http://www.youtube.com/watch?v=U-XuCRO1Ecw,

16. "Randy Pausch Last Lecture: Achieving Your Childhood Dreams," 2007, http://www.youtube.com/watch?v=ji5_MqicxSo.

17. "Randy Pausch Inspires Graduates," 2008, http://www.youtube.com/watch?v=RcYv5x6gZTA.

18. Other lecturers in the Journeys series also followed this format, although they spoke to much smaller audiences.

19. The video does not indicate if royalties were paid for recording and distributing this song, which has been at the center of several copyright battles. As to the "Last Lecture" itself, on Pausch's website he indicates that he will eventually seek a Creative Commons license, although at the time of his death it was still posted with a traditional copyright mark.

20. Dean Richard Miksad, who is identifiable, despite his blacked-out visage, stepped down in 2004.

21. Gabriel Robins, "'Time Management' by Randy Pausch," November 2007, http://www.youtube.com/watch?v=blaK_tB_KQA.

22. Alexis Madrigal, "Mourning the Internet Famous: Randy Pausch's Distributed Funeral." *Wired*, July 29, 2008, http://www.wired.com/culture/lifestyle/news/2008/07/distributed_funeral.

23. Geert Lovink, *Zero Comments: Blogging and Critical Internet Culture* (New York: Routledge, 2007).

24. Stanley Eugene Fish, *There's No Such Thing as Free Speech* (Durham, NC: Duke University Press, 1994).

25. wonderingmind42, "The Most IMPORTANT Video You'll Ever See," June 16, 2007, http://www.youtube.com/watch?v=F-QA2rkpBSY.

26. Michael J. De La Merced, "A Student's Video Résumé Gets Attention (Some of It Unwanted)," *New York Times*, October 21, 2006, Business sec., http://www.nytimes.com/2006/10/21/business/21bank.html.

27. Ben McGrath, "Top This: Aleksey the Great," *New Yorker*, October 23, 2006, http://www.newyorker.com/archive/2006/10/23/061023ta_talk_mcgrath.

28. Aleksey Vayner, "Aleksey Vayner Discusses Impossible Is Nothing Video at Harvard/MIT Conference," http://www.youtube.com/watch?v=MnRhqWWHsRw April 29, 2011.

29. "'Do Not, Anyone, Sell This Idiot ANY Pills!' The Desperate Last Messages to Former Yale Student Infamous for 'Impossible Is Nothing' Résumé Who Is Reportedly 'Dead at Age 29 from an Overdose,'" *Mail Online*, http://www.dailymail.co.uk/news/article-2267861/Aleksey-Vayner-Impossible-Nothing-r-sum-star-dead-overdose.html.

4 From Reality TV to the Research University: Coursecasting and Pedagogical Drama

1. The concept for this course design came from discussions with Ian Bogost and N. Katherine Hayles about how to teach the writing of electronic literature, and I am grateful to them for their many suggestions about paths to pedagogical implementa-

tion, which I still use with my digital poetics students. For more about electronic literature, see materials from the Electronic Literature Organization at http://eliterature.org/. For more about the course, see the syllabus at http://losh.ucsd.edu/courses/poetics.html.

2. "MIT Professor Makes Physics Fun on YouTube," Early Show, CBS, December 20, 2007, http://www.videowired.com/video/852034825/.

3. "Physics I: Classical Mechanics," MIT OpenCourseWare, fall 1999, http://ocw.mit.edu/OcwWeb/Physics/8-01Physics-IFall1999/CourseHome/index.htm.

4. "Walter Lewin Promo," December 20, 2007, http://www.youtube.com/watch?v=7Zc9Nuoe2Ow.

5. Sara Rimer, "At 71, Physics Professor Is a Web Star," *New York Times*, December 19, 2007, http://www.nytimes.com/2007/12/19/education/19physics.html.

6. "iTunes U," http://www.apple.com/education/itunesu_mobilelearning/itunesu.html.

7. Brock Read, "How to Podcast Campus Lectures," *Chronicle of Higher Education*, January 26, 2007, Technology sec., http://chronicle.com/article/How-to-Podcast-Campus-Lectures/25128.

8. Urs Gasser, "iTunes: How Copyright, Contract, and Technology Shape the Business of Digital Media—A Case Study," *SSRN eLibrary* (June 2004), http://papers.ssrn.com/sol3/papers.cfm?abstract_id=556802.

9. "Play: BARACK PAPER SCISSORS," November 7, 2008, http://www.youtube.com/watch?v=l2mcdS6ioo8.

10. Stephen Ramsay, Faculty Development Leave Letter to Susan Belasco, July 12, 2012.

11. "I Need an A (R-Rated Version)," December 14, 2010, http://www.youtube.com/watch?v=6lfUEefilVI.

12. "One Professor's Fantasy," October 23, 2010, http://www.youtube.com/watch?v=qeSdC7lbAlA.

13. "So You Want to Get a PhD in the Humanities," October 25, 2010, http://www.youtube.com/watch?v=obTNwPJvOI8.

14. Alex Usher, "Coursera Jumps the Shark," Higher Education Strategy Associates, May 31, 2013. http://higheredstrategy.com/coursera-jumps-the-shark/.

15. For example, Jonathan Alexander and I have asserted that revelatory moments in reality television shows and public service announcements may play a role in composing some coming-out videos by queer youth on YouTube. See Christopher Pullen and Margaret Cooper, *LGBT Identity and Online New Media* (New York: Routledge, 2010).

16. Mary Madden, "Podcast Downloading," Pew Internet Project Data Memo, November 2006, http://www.pewinternet.org/Reports/2006/Podcast-Downloading.aspx.

17. "19% of Internet Users Have Downloaded a Podcast," Pew Internet & American Life Project, n.d., http://pewinternet.org/Reports/2008/Podcast-Downloading-2008/Data-Memo.aspx.

18. Richard Leppert, "The Social Discipline of Listening," in *Aural Cultures,* ed. Jim Drobnick (Toronto: XYZ Books, 2004), 27.

19. Lisa Gitelman, *Always, Already New: Media, History, and the Data of Culture* (Cambridge, MA: MIT Press, 2006).

20. Mizuko Ito, Daisuke Okabe, and Misa Matsuda, *Personal, Portable, Pedestrian* (Cambridge, MA: MIT Press, 2005).

21. Michael Heim, *Virtual Realism* (New York: Oxford University Press, 1998), 84.

22. Patricia Meyer Spacks, *Gossip* (New York: Knopf, 1985).

23. See, e.g., in the field of composition, Mina P. Shaughnessy, *Errors and Expectations: A Guide for the Teacher of Basic Writing* (New York: Oxford University Press, 1977).

24. Heather Hendershot, "Making It Work on *The Simple Life* and *Project Runway,*" in *Reality TV: Remaking Television Culture,* ed. Susan Murray (New York: New York University Press, 2009), 243–259.

25. Toby Miller, *Blow Up the Humanities* (Philadelphia: Temple University Press, 2012), 25–27.

26. Rebecca Weeks, "Runaway Ratings for Project Runway," *iMedia Connection,* March 21, 2006, http://www.imediaconnection.com/content/8729.asp.

27. Ibid.

28. Ibid.

29. Hillel Schwartz, *The Culture of the Copy: Striking Likenesses, Unreasonable Facsimiles* (New York: Zone Books, 1996).

30. Henry Jenkins, *Convergence Culture: Where Old and New Media Collide* (New York: New York University Press, 2006).

31. Ibid.

32. Walter Ong, *Orality and Literacy: The Technologizing of the Word* (New York: Methuen, 1982), 133–134.

33. Henry Jenkins, *Spreadable Media: Creating Value and Meaning in a Networked Culture* (New York: New York University Press, 2013).

34. *Jill Bolte Taylor's Stroke of Insight*, TED.com, n.d., http://www.ted.com/talks/jill_bolte_taylor_s_powerful_stroke_of_insight.html.

35. *Lies, Damned Lies, and Statistics (about TEDTalks)*, TED.com, n.d., http://www.ted.com/talks/lies_damned_lies_and_statistics_about_tedtalks.html.

36. Nathan Heller, "Listen and Learn." *New Yorker*, July 9, 2012.

37. Benjamin Wallace, "Those Fabulous Confabs." *New York Magazine*, February 26, 2012.

38. Step-by-step Photoshop videos can also be a source of social commentary, satire, parody, and play. See, for example, the chapter on Photoshop in "Submit and Render" that describes the public response to the Dove "Evolution of Beauty" commercial in Elizabeth Losh, *Virtualpolitik: An Electronic History of Government Media-making in a Time of War, Scandal, Disaster, Miscommunication, and Mistakes* (Cambridge, MA: MIT Press, 2009).

39. "dekePod: 101 Photoshop Tips in Five Minutes," June 23, 2008. http://www.youtube.com/watch?v=h6flegTtolg.

40. However, as Marjorie Curry Woods points out, instruction in medieval rhetoric also included exercises devoted not to copying formal elements but to evoking the feelings of a specific character with a performative, rather than technical, pedagogical philosophy at work. For example, rhetorical training that incorporated Ethopoeia or Impersonation might include popular exercises called progymnasmata that could involve transgender identification. For example, Augustine describes inhabiting the spirit of Juno or Dido as a schoolboy, rendering his own version of their speeches. See Margaret Curry Woods, "Weeping for Dido: Epilogue on a Premodern Rhetorical Exercise in the Postmodern Classroom," in *Latin Grammar and Rhetoric: From Classical Theory to Medieval Practice*, ed. Carol Dana Lanham (New York: Continuum, 2002).

41. Alison King, "From Sage on the Stage to Guide on the Side," *College Teaching* 41, no. 1 (1993): 30–35.

42. Gregory Clark and S. Michael Halloran, *Oratorical Culture in Nineteenth-Century America: Transformations in the Theory and Practice of Rhetoric* (Carbondale: Southern Illinois University Press, 1993).

5 The Rhetoric of the Open Courseware Movement

1. Tamar Lewin, "Consortium of Colleges Takes Online Education to New Level," *New York Times*, July 17, 2012, http://www.nytimes.com/2012/07/17/education/consortium-of-colleges-takes-online-education-to-new-level.html.

2. See http://www.khanacademy.org.

3. "KA Lite, an Offline Version of Khan Academy," http://kalite.learningequality.org/.

4. Lewin, "Consortium of Colleges Takes Online Education to New Level."

5. "Introduction to Artificial Intelligence," October—December 2011, https://www.ai-class.com/.

6. Norvig is well known to those who study digital rhetoric as the author of the "Gettysburg PowerPoint Presentation" that mocks the popular Microsoft slideware product, and he has more recently entered the political arena in advertisements for Patriotic Millionaires for Fiscal Strength, in which he asks to pay more taxes to fund government programs.

7. "Unit 0w, 1 Introduction," October 5, 2011, http://www.youtube.com/watch?v=BnIJ7Ba5Sr4.

8. "Office Hours Part 1," November 9, 2011, http://www.youtube.com/watch?v=pIfrmdM0Ht0.

9. See http://www.udacity.com/.

10. Jeffrey R. Young, "Inside the Coursera Contract: How an Upstart Company Might Profit from Free Courses," *Chronicle of Higher Education*, July 19, 2012, http://chronicle.com/article/How-an-Upstart-Company-Might/133065/.

11. See http://see.stanford.edu/.

12. See https://www.edx.org/.

13. "The U. of Michigan's Contract with Coursera," *Chronicle of Higher Education*, July 19, 2012, http://chronicle.com/article/Document-Examine-the-U-of/133063/.

14. "Coursera FAQs," http://provost.upenn.edu/initiatives/openlearning/coursera faqs.

15. *Daphne Koller: What We're Learning from Online Education*, TED.com, http://www.ted.com/talks/daphne_koller_what_we_re_learning_from_online_education.html.

16. Irene Ogrizek, "Daphne Koller and the Problem with Coursera," *Irene Ogrizek*, May 18, 2013, http://ireneogrizek.ca/2013/05/18/8932/.

17. Daphne Koller, "The Online Revolution: Education for Everyone," University of California, San Diego, February 13, 2013.

18. Anant Agarwal, "Online Universities: It's Time for Teachers to Join the Revolution," *Guardian*, June 15, 2013, http://www.guardian.co.uk/education/2013/jun/15/university-education-online-mooc.

19. "Picture of the Day: Udacity Students in West Africa," http://blog.udacity.com/2012/08/picture-of-day-udacity-students-in-west.html.

20. Rayvon Fouché, "From Black Inventors to One Laptop per Child: Exporting a Racial Politics of Technology," in *Race after the Internet*, ed. Lisa Nakamura and Peter Chow-White (New York: Routledge, 2012).

21. Nakamura and Chow-White, *Race after the Internet*, 10.

22. Siva Vaidhyanathan, *The Googlization of Everything (and Why We Should Worry)* (Berkeley: University of California Press, 2011).

23. "Welcome to Course Builder!," https://code.google.com/p/course-builder/.

24. At the time of this writing, Bogost's own institution had developed an online computer science degree program through a partnership with Udacity and AT&T that had generated "significant internal disagreements." Faculty were unhappy with what they perceived to be a deal that would create entire classes of nonfaculty positions to be staffed by instructional technologists. Some in the tech industry described it as a "$7000 polyester masters" degree. For more, see Ry Rivard, "Documents Shed Light on Details of Georgia Tech-Udacity Deal," *Inside Higher Ed*, May 28, 2013, http://www.insidehighered.com/news/2013/05/28/documents-shed-light-details-georgia-tech-udacity-deal, and Robert Cringely, "Georgia Tech's $7000 Polyester Masters in Computer Science," *I, Cringely*, July 10, 2013, http://www.cringely.com/2013/07/10/georgia-techs-7000-polyester-masters-in-computer-science/.

25. "MOOCs and the Future of the Humanities: A Roundtable (Part 1) by Ian Bogost, Cathy N. Davidson, Al Filreis, and Ray Schroeder," *Los Angeles Review of Books*, June 14, 2013, http://www.lareviewofbooks.org/article.php?id=1757&fulltext=1.

26. Fred Turner, "How Digital Technology Found Utopian Ideology: Lessons From the First Hackers' Conference," in *Critical Cyberculture Studies*, ed. David Silver (New York: New York University Press, 2006).

27. The lectures of this popular large-enrollment Harvard course have been a campus staple for decades and were among the first to be marketed online to the public by the university. The author had been a student in the traditional face-to-face learning environment in Sanders Theater, but she was invited to retake the course online with other alumni. The videos for Sandel's lectures also reflect his partnership with public broadcaster WGBH television and sponsorship by the POM Wonderful juice company. The website justiceharvard.org archives the lectures and facilitates "community" around the course.

28. "An Open Letter to Professor Michael Sandel from the Philosophy Department at San Jose State U," *Chronicle of Higher Education*, May 2, 2013, http://chronicle.com/article/The-Document-Open-Letter-From/138937/.

29. "Michael Sandel Responds," *Chronicle of Higher Education*, May 2, 2013, http://chronicle.com/article/Michael-Sandel-Responds/139021/.

30. "SB 520—AMENDED," http://leginfo.legislature.ca.gov/faces/billNavClient.xhtml?bill_id=201320140SB520.

31. "An Open Letter to Members of the Academic Senate about SB 520," http://senate.universityofcalifornia.edu/RP_BJ2AllSenate_SB520_031513.pdf.

32. Andrew Rice, "What the Failed Removal of UVA President Teresa Sullivan Means for Higher Education," *New York Times*, September 11, 2012, http://www.nytimes.com/2012/09/16/magazine/teresa-sullivan-uva-ouster.html.

33. Siva Vaidhyanathan, "Strategic Mumblespeak," *Slate*, June 15, 2012, http://www.slate.com/articles/news_and_politics/hey_wait_a_minute/2012/06/teresa_sullivan_fired_from_uva_what_happens_when_universities_are_run_by_robber_barons_.single.html.

34. Clayton M. Christensen and Henry J. Eyring, *The Innovative University: Changing the DNA of Higher Education from the Inside Out* (San Francisco: Jossey-Bass, 2011), 20.

35. Adele Clarke and Joan H. Fujimura, *The Right Tools for the Job: At Work in Twentieth-Century Life Sciences* (Princeton, NJ: Princeton University Press, 1992).

36. See, e.g., Kalindi Vora, "Limits of 'Labor': Accounting for Affect and the Biological in Transnational Surrogacy and Service Work," *South Atlantic Quarterly* 111, no. 4 (fall 2012): 681–700, and Chikako Takeshita, *The Global Biopolitics of the IUD: How Science Constructs Contraceptive Users and Women's Bodies* (Cambridge, MA: MIT Press, 2012).

37. See the forthcoming *The Digital Classroom in the Time of Wikipedia*, ed. Nishant Shah.

38. It is striking, in fact, to note that Wikipedia only appears in the pages of *The Innovative University* as the source for a note about the past history of "online learning's technological immaturity" that cites statistics about Internet access in the 1990s. Christensen and Eyring, *The Innovative University*, 209.

39. Laura Pappano, "Massive Open Online Courses Are Multiplying at a Rapid Pace," *New York Times*, November 2, 2012, http://www.nytimes.com/2012/11/04/education/edlife/massive-open-online-courses-are-multiplying-at-a-rapid-pace.html.

40. "What You Need to Know about MOOCs," *Chronicle of Higher Education*, 2013, http://chronicle.com/article/What-You-Need-to-Know-About/133475/.

41. Thomas L. Friedman, "Revolution Hits the Universities," *New York Times*, January 26, 2013, http://www.nytimes.com/2013/01/27/opinion/sunday/friedman-revolution-hits-the-universities.html, and Thomas L. Friedman, "Come the Revolution," *New York Times*, May 15, 2012, http://www.nytimes.com/2012/05/16/opinion/friedman-come-the-revolution.html.

42. See http://www.moocresearch.com.

43. Coursera Staff, "Technical Issues," September 15, 2012.

44. Lance Gordon, "UC Irvine Professor Quits Midway through Online Coursera Class," *Los Angeles Times*, February 18, 2013, http://latimesblogs.latimes.com/lanow/2013/02/uc-irvine-business-professor-stops-teaching-midway-in-online-coursera-class.html.

45. "Coursera Forced to Call Off a MOOC amid Complaints about the Course," Inside Higher Ed," February 4, 2013, http://www.insidehighered.com/news/2013/02/04/coursera-forced-call-mooc-amid-complaints-about-course.

46. "How NOT to Design a MOOC: The Disaster at Coursera and How to Fix It," 2013, *Online Learning Insights*, http://onlinelearninginsights.wordpress.com/2013/02/01/how-not-to-design-a-mooc-the-disaster-at-coursera-and-how-to-fix-it/.

47. "Fundamentals of Online Education: Planning and Application with Fatimah Wirth," July 16, 2012, http://www.youtube.com/watch?v=-MlSs0gASvU.

48. "My First MOOC: Online Class about How to Create Online Classes Failed Miserably," Digital/Edu, 2013, http://digital.hechingerreport.org/content/my-first-mooc-crashed_371/.

49. "Coursera Forced to Call Off a MOOC amid Complaints about the Course," http://www.insidehighered.com/news/2013/02/04/coursera-forced-call-mooc-amid-complaints-about-course.

50. Alex Reid, "Why Utopian/Dystopian Thinking Is Wrong-Headed #edcmooc," Digital Digs, January 29, 2013, http://www.alex-reid.net/2013/01/why-utopiandystopian-thinking-is-wrong-headed-edcmooc.html.

51. Sandra K. Milligan, "Better than a Tarantino Movie: Raw Peer Assessment in #edcmooc," 2013, http://sandrakmilligan.wordpress.com/2013/03/05/better-than-a-tarantino-movie-raw-peer-assessment-in-a-mooc/.

52. Steve Krause, "E-Learning and Digital Cultures Ends with a Meh #edcmooc," March 6, 2013,http://stevendkrause.com/2013/03/06/e-learning-and-digital-cultures-ends-with-a-meh-edcmooc/.

53. Csikszentmihalyi Mihaly, *Flow: The Psychology of Optimal Experience* (New York: Harper & Row, 1990).

54. Derek Lomas, Kishan Patel, Jodi L. Forlizzi, and Kenneth R. Koedinger, "Optimizing Challenge in an Educational Game Using Large-Scale Design Experiments," in *Proceedings of the SIGCHI Conference on Human Factors in Computing Systems* (CHI '13) (New York: ACM Press, 2013), 89–98.

55. Daphne Koller, Andrew Ng, Chong Do, and Chen Zhanghao, "Retention and Intention in Massive Open Online Courses," *Educause Review Online*, June 3, 2013,

http://www.educause.edu/ero/article/retention-and-intention-massive-open-online-courses.

56. Glenda M. Crosling and Liz Thomas, *Improving Student Retention in Higher Education: The Role of Teaching and Learning* (New York: Routledge, 2008).

57. See http://connectedlearning.tv/.

58. Ibid.

59. Ibid.

60. Audrey Watters, "A Bill of Rights and Principles for Learning in the Digital Age," https://github.com/audreywatters/learnersrights/blob/master/bill_of_rights.md.

61. See work of the Stanford Lytics Lab, on the effects of social cues in video instruction on learning and persistence in MOOCs by Rene Kizilcec, Emily Schneider, and Dan McFarland.

62. Tamar Lewin, "Colleges Turn to Crowd-Sourcing Courses," *New York Times*, November 19, 2012, http://www.nytimes.com/2012/11/20/education/colleges-turn-to-crowd-sourcing-courses.html.

63. "Duke Students Use Coursera in 'Flipped Classroom,'" *Duke Chronicle*, http://www.dukechronicle.com/articles/2013/03/05/duke-students-use-coursera-flipped-classroom.

64. Marc Parry, "A Star Professor Defects—at Least for Now," *Chronicle of Higher Education*, September 3, 2013, http://chronicle.com/article/A-MOOC-Star-Defects-at-Least/141331/.

65. Theophrastus, "The Problems with MOOCs 1: Robo-Essay Grading," *BLT*, April 5, 2013, http://bltnotjustasandwich.com/2013/04/05/the-problems-with-moocs-1-robo-essay-grading/.

66. Les Perelman, "Critique (Ver. 3.4) of Mark D. Shermis and Ben Hammer, 'Contrasting State of the Art Automated Scoring of Essays: Analysis,'" http://graphics8.nytimes.com/packages/pdf/science/Critique_of_Shermis.pdf.

67. For more about the self-help movement, see Micki McGee, *Self-Help, Inc.: Makeover Culture in American Life* (Oxford: Oxford University Press, 2005).

68. Lynee Lewis Gaillet, "An Historical Perspective on Collaborative Learning," *Journal of Advanced Composition* 14, no. 1 (1994): 93–110.

69. Ayhan Aytes, "Return of the Crowds: Mechanical Turk and Neoliberal States of Exception," in *Digital Labor: The Internet as Playground and Factory*, ed. Trebor Scholz (New York: Routledge, 2013).

70. Cathy Davidson, "How to Crowdsource Grading," HASTAC, 2013, http://hastac.org/blogs/cathy-davidson/how-crowdsource-grading.

71. Keith Devlin, "How Are MOOCs Organized?," MOOCtalk, February 23, 2013, http://mooctalk.org/tag/coursera/.

72. Questions around the subjectivity of identifying errors have been some of the most vexing in composition studies. The evolution of error-finding practices among communities of instructors has a long rhetorical history. See Tracy Santa, *Dead Letters: Error in Composition, 1873–2004* (Cresskill, NJ: Hampton Press, 2008). Furthermore, even expert readers may locate errors differently. See Mina P. Shaughnessy, *Errors and Expectations: A Guide for the Teacher of Basic Writing* (New York: Oxford University Press, 1977).

73. Elizabeth Losh, "Virtualpolitik: Obstacles to Building Virtual Communities in Traditional Institutions of Knowledge. Research & Occasional Paper Series: CSHE.9.05," online submission, June 2005, http://eric.ed.gov/?id=ED492224.

74. "Calibrated Peer Review: Overview," http://cpr.molsci.ucla.edu/Overview.aspx.

75. "CPR Publications," http://www.chem.ucla.edu/dept/Faculty/russell/CPR%20 Publications.pdf.

76. Arlene Russell, Calibrated Peer Review Presentation, March 8, 2013, University of California, San Diego.

77. Nancy Sommers, "Responding to Student Writing," *College Composition and Communication* 33, no. 2 (May 1982): 148–156.

78. Jeremy N. Bailenson, Nick Yee, Dan Merget, and Ralph Schroeder, "The Effect of Behavioral Realism and Form Realism of Real-Time Avatar Faces on Verbal Disclosure, Nonverbal Disclosure, Emotion Recognition, and Copresence in Dyadic Interaction," *Presence* 15, no. 4 (August 2006): 359–372.

79. "MIT Media Lab: Affective Computing Group," http://affect.media.mit.edu/.

80. Scout Finch, "Flipped Class Hate Tweets," n.d., https://docs.google.com/file/d/0 B7PbBPHGXloTTTFBY050Tld0SWs.

81. Kamala Visweswaran, *Fictions of Feminist Ethnography* (Minneapolis: University of Minnesota Press, 1994).

82. Tom Boellstorff, Bonnie Nardi, Celia Pearce, and T. L. Taylor, *Ethnography and Virtual Worlds: A Handbook of Method* (Princeton, NJ: Princeton University Press, 2012).

83. Sarah Pink, *Doing Sensory Ethnography* (Los Angeles: SAGE, 2009).

84. "Reclaim Open Learning," http://open.media.mit.edu/.

85. Anya Kamenetz, "Can We Move Beyond the MOOC to Reclaim Open Learning?," *Huffington Post,* June 16, 2013, http://www.huffingtonpost.com/anya-kamenetz/can-we-move-beyond-the-mooc_b_3451418.html.

86. "Free software manifesto," http://www.gnu.org/philosophy/free-sw.html.

87. Audrey Watters, "[Expletive Deleted] Ed-Tech #Edinnovation," *Hack Education*, May 4, 2013, http://www.hackeducation.com/2013/05/04/ed-tech-argo-f-k-yourself/.

88. David Wiley, "IPT 692R: Introduction to Open Education," http://blip.tv/introduction-to-open-education/lecture-01-history-1653975.

89. "HILT 47 Shigehisa Kuriyama," May 7, 2013, http://www.youtube.com/watch?v=tL8yM2A7o2U.

90. James Paul Gee, *What Video Games Have to Teach Us about Learning and Literacy*, rev. and updated ed. (New York: Palgrave, 2007).

91. José Antonio Bowen, *Teaching Naked: How Moving Technology Out of Your College Classroom Will Improve Student Learning* (San Francisco, CA: Wiley, 2012).

92. David Wiley, "More on MOOCs and Being Awesome Instead," *Iterating Toward Openness*, May 24, 2013, http://opencontent.org/blog/archives/2879.

93. David Wiley, "OpenCourseWare Deleted Scenes," http://www.opencontent.org/future/.

94. Toru Iiyoshi, M. S. Vijay Kumar, and Carnegie Foundation for the Advancement of Teaching, *Opening Up Education: The Collective Advancement of Education through Open Technology, Open Content, and Open Knowledge* (Cambridge, MA: MIT Press, 2008), 259.

95. Ibid., 197.

96. Ibid., 105.

97. Ibid., 107.

98. "Lovemarks: Harvard University (Nomination)," n.d., http://www.lovemarks.com/index.php?pageID=20015&lovemarkid=2801.

99. William Mitchell, *Placing Words: Symbols, Space, and the City* (Cambridge, MA: MIT Press, 2005).

100. Thomas Friedman actually praises these lectures in "The Professors' Big Stage," *New York Times*, March 5, 2013, Opinion sec., http://www.nytimes.com/2013/03/06/opinion/friedman-the-professors-big-stage.html.

101. See, e.g., Kathleen Fitzpatrick, *Planned Obsolescence: Publishing, Technology, and the Future of the Academy* (New York: New York University Press, 2011), which also argues for systems of open peer review quite different from conventional blind peer review practices.

102. John Willinsky, *The Access Principle: The Case for Open Access to Research and Scholarship* (Cambridge, MA: MIT Press, 2006), 144.

103. Ibid., 150.

104. Stefan Baldi, Hauke Heier, and Fabian Stanzick, "Open Courseware vs. Open Source Software—A Critical Comparison," *ECIS 2002 Proceedings*, January 1, 2002, http://aisel.aisnet.org/ecis2002/146.

105. Jonathan Haber, "xMOOC vs. cMOOC—MOOC Pedagogy," *Degree of Freedom*, April 29, 2013, http://degreeoffreedom.org/xmooc-vs-cmooc/.

106. Will Oremus, "The New Public Ivies," *Slate*, July 18, 2012, http://www.slate.com/articles/technology/future_tense/2012/07/coursera_udacity_edx_will_free_online_ivy_league_courses_end_the_era_of_expensive_higher_ed_.html.

107. Anya Kamenetz, "Can We Move Beyond the MOOC to Reclaim Open Learning?," *Huffington Post*, June 16, 2013, http://www.huffingtonpost.com/anya-kamenetz/can-we-move-beyond-the-mooc_b_3451418.html.

108. Hakim Bey, *TAZ: The Temporary Autonomous Zone* (Seattle, WA: Pacific Publishing Studio, 2011).

109. April Laskey Aerni and KimMarie McGoldrick, *Valuing Us All: Feminist Pedagogy and Economics* (Ann Arbor: University of Michigan Press, 1999).

110. See Amy Earhart, "Recovering the Recovered Text: Diversity, Canon Building, and Digital Studies," http://lecture2go.uni-hamburg.de/konferenzen/-/k/13976.

111. "Transforming Higher Education with Distributed Open Collaborative Courses (DOCCs): Feminist Pedagogies and Networked Learning," http://femtechnet.newschool.edu/femtechnet-whitepaper/.

112. bell hooks, *Teaching to Transgress: Education as the Practice of Freedom* (New York: Routledge, 1994).

113. Elizabeth G. Peck and JoAnna Stephens Mink, *Common Ground Feminist Collaboration in the Academy* (Albany: SUNY Press, 1998).

114. Carolyn M. Shrewsbury, "What Is Feminist Pedagogy?" *Women's Studies Quarterly* 15, no. 3–4 (1987): 6–14.

115. Megan Boler, *Feeling Power: Emotions and Education* (New York: Routledge, 1999).

116. Ellen Messer-Davidow, *Disciplining Feminism: From Social Activism to Academic Discourse* (Durham: Duke University Press, 2002).

117. "What Is Feminist IT?," http://feministit.ucsd.edu/about/.

118. Alexandra Juhasz, e-mail of September 12, 2012.

119. The descriptor "massive" was once celebrated because of its association with networked multiplayer games that seemed to undermine impersonal hegemonic

authority. See "MASSIVE: The Future of Networked Multiplayer Games," http://www.isr.uci.edu/events/massive/.

120. Armando Fox, "Viewpoint: MOOC Mythbusting," *Armando Fox*, January 30, 2013, http://www.armandofox.com/geek/2012/11/viewpoint-moocs-can-strengthen-academia/.

121. Daniel M. Russell, Scott Klemmer, Armando Fox, Celine Latulipe, Mitchell Duneier, and Elizabeth Losh, "Will Massive Online Open Courses (MOOCs) Change Education?," in *CHI '13 Extended Abstracts on Human Factors in Computing Systems* (New York: ACM Press, 2013), 2395–2398.

122. Armando Fox, "What Was It Like to Teach a MOOC," *Armando Fox*, January 14, 2013, http://www.armandofox.com/geek/2013/01/what-was-it-like-to-teach-a-mooc/.

123. For more, see "FAQ for FemTechNet," *Fembot Collective*, http://fembotcollective.org/femtechnet/faq-for-femtechnet/ and "About FemTechNet," FemTechNet Commons, http://femtechnet.newschool.edu/the-network/.

124. Susan Leigh Star and James R. Griesemer, "Institutional Ecology, 'Translations' and Boundary Objects: Amateurs and Professionals in Berkeley's Museum of Vertebrate Zoology, 1907–39," *Social Studies of Science* 19, no. 3 (August 1, 1989): 393.

125. Ibid.

126. Susan Leigh Star, "This Is Not a Boundary Object: Reflections on the Origin of a Concept," *Science, Technology & Human Values* 35, no. 5 (September 1, 2010): 601–617, and Geoffrey C. Bowker and Susan Leigh Star, *Sorting Things Out Classification and Its Consequences* (Cambridge, MA: MIT Press, 1999).

127. "FAQ for FemTechNet," *Fembot Collective*, http://fembotcollective.org/femtechnet/faq-for-femtechnet/.

128. Elizabeth Losh, "Learning from Failure: Feminist Dialogues on Technology, Part II," *DML Central*, August 9, 2012, http://dmlcentral.net/blog/liz-losh/learning-failure-feminist-dialogues-technology-part-ii.

129. Elizabeth Losh, "Bodies in Classrooms: Feminist Dialogues on Technology, Part I," *DML Central*, n.d., http://dmlcentral.net/blog/liz-losh/bodies-classrooms-feminist-dialogues-technology-part-i.

130. Forum topic 1373.

6 Honor Coding: Plagiarism Software and Educational Opportunism

1. "WeTakeYourClass.com (Pay Someone to Take My Online Class)," October 19, 2012, http://www.youtube.com/watch?v=qp4TwDwE8GM.

2. "WeTakeYourClass.com—About Us," 2013, http://www.wetakeyourclass.com/about-us.html.

3. "Hire a Professional to Take My Online Class? Noneedtostudy.com," January 20, 2012, http://www.youtube.com/watch?v=gzGQ7wnOjqM&.

4. For a pre-Internet account of ghost writing for a paper mill, see Abigail Witherspoon, "This Pen for Hire: On Grinding Out Papers for College Students," *Harper's* (June 1995): 49–57.

5. Ed Dante, "The Shadow Scholar," *Chronicle of Higher Education*, November 12, 2010, http://chronicle.com/article/The-Shadow-Scholar/125329/.

6. Ibid.

7. Jeffrey R. Young, "Online Classes See Cheating Go High-Tech," *Chronicle of Higher Education*, June 3, 2012, http://chronicle.com/article/Cheating-Goes-High-Tech/132093/.

8. Karen Thomas, "Colleges Clamp Down on Cheaters," *USA Today*, June 20, 2001, http://usatoday30.usatoday.com/news/nation/june01/2001-06-11-cheaters-sidebar.htm.

9. See http://www3.dbu.edu/library/faculty/turnitin_information.asp.

10. Anita Kumar, "State: High-Tech Sleuthing Catches College Cheats," *St. Petersburg Times Online*, August 31, 2003, http://www.sptimes.com/2003/08/31/news_pf/State/High_tech_sleuthing_c.shtml.

11. Elizabeth Murphy, "Plagiarism Software WriteCheck Troubles Some Educators," *USA Today*, September 9, 2011, http://www.usatoday.com/news/education/story/2011-09-09/college-cheating-plagiarism/50338736/1.

12. Charles Bazerman, "Paying the Rent: Languaging Particularity and Novelty," paper presented at the Originality, Imitation, and Plagiarism Conference, University of Michigan, August 25, 2005.

13. Tarleton Gillespie, Pablo J. Boczkowski, and Kirsten A. Foot, *Media Technologies: Essays on Communication, Materiality, and Society* (Cambridge, MA: MIT Press, 2014).

14. Manuel Castells, *The Rise of the Network Society* (Malden, MA: Blackwell, 1996).

15. David Kay, August 6, 2002, e-mail.

16. Sharon Block, quoted in Paula Murphy, "New Technology Update: Electronic Plagiarism Detection," *TLtC News*, UC Center for Teaching and Learning with Technology (May–June 2004). The metaphor of "false positives" from Turnitin also appears in Adam L. Penenberg, "Me Against My Students: How I Use the Internet to Combat Plagiarists, Fabulists, and Cheaters," *Slate,* October 3, 2005, http://www.slate.com/id/2127365/?nav=tap3.

17. Barbara Ehrenreich, *Nickel and Dimed: On (Not) Getting By in America* (New York: Henry Holt, 2001).

18. In a post–September 11th society, one in which the color coding by the Office of Homeland Security uses a similar index, it is hard to see this system as politically neutral, even though the index predates the terrorist threat level system.

19. "Turnitin FAQs," http://turnitin.com/en_us/features/faqs.

20. Although, considering that Turnitin began with a peer-review program, it is questionable to see these activities as new.

21. Turnitin, "Legal Document," http://www.turnitin.com/static/legal/legal_document.html (accessed September 23, 2005; no longer available online).

22. Turnitin, "Technology FAQ," https://www.turnitin.com/static/faqs/technology_faq.html (accessed September 23, 2005; no longer available online).

23. The Center for Academic Integrity, "How to Get Started," http://www.academicintegrity.org/resources_inst.asp (accessed September 23, 2005; no longer available online).

24. Quoted in "A Focus on Academic Integrity: Code Found Not to Be a Part of Campus Culture," *Duke News and Communications*, March 9, 2001, http://today.duke.edu/2001/03/integrity309.html.

25. Quoted in Mary Clarke-Pearson, "Download. Steal. Copy. Cheating at the University," *Daily Pennsylvanian*, November 27, 2001, http://www.dailypennsylvanian.com/vnews/display.v/ART/2001/11/27/3c03502bad345?in_archive=1.

26. Quoted in Grace Lee, "Plagiarism 101," *Read Me: New Media. Net Culture. Now*, http://journalism.nyu.edu/pubzone/ReadMe/article.php?id=441.

27. The Center for Academic Integrity, "How to Get Started," http://www.academicintegrity.org/resources_inst.asp (accessed September 23, 2005; no longer available online).

28. Kelly Ritter, "The Economics of Authorship: Online Paper Mills, Student Writers, and First-Year Composition," *CCC* 56, no. 4 (June 2005): 625.

29. Rebecca Moore Howard, "Plagiarisms, Authorships, and the Academic Death Penalty," *College English* 57, no. 7 (November 1995): 788–806.

30. James R. Purdy, "Calling off the Hounds: Technology and the Visibility of Plagiarism," *Pedagogy* 5, no. 2 (2005): 277.

31. Jeffrey R. Young, "The Cat and Mouse Game of Plagiarism Detection," *Chronicle of Higher Education*, http://chronicle.com/free/v47/i43/43a02601.htm.

32. Peter Levin, "Beat the Witch-Hunt! Peter Levin's Guide to Avoiding and Rebutting Accusations of Plagiarism, for Conscientious Students," 2003, http://student-friendly-guides.com/wp-content/uploads/Beat-the-Witch-hunt.pdf.

33. Advocates for calling plagiarism "fraud" rather than "theft" do not necessarily question the analogy of intellectual property to tangible goods, as I do in the subtext of this chapter.

34. "Highway robbery" could be seen in two different contexts at http://www.fno.org/may98/cov98may.html and http://www.camlang.com/sp005.htm (accessed September 23, 2005; no longer available online).

35. Candace Spigelman, *Across Property Lines: Textual Ownership in Writing Groups*, Studies in Writing and Rhetoric (Carbondale: Southern Illinois University Press, 2000), 23.

36. American Library Association, "Information Literacy for Faculty and Administrators," http://www.ala.org/acrl/issues/infolit/overview/faculty/faculty.

37. "Faculty Perceptions of Plagiarism," *Journal of College and Character* 2 (2005), http://www.degruyter.com/view/j/jcc.2005.6.2/jcc.2005.6.2.1416/jcc.2005.6.2.1416.xml.

38. Ritter, "The Economics of Authorship," 626–627.

39. Purdy, "Calling Off the Hounds," 275–296.

40. The Center for Academic Integrity, "How to Get Started," http://www.academicintegrity.org/resources_inst.asp (accessed September 23, 2005; no longer available online).

41. Ritter, "The Economics of Authorship," 627.

42. Sara Rimer, "A Campus Fad That's Being Copied: Internet Plagiarism," *New York Times*, September 23, 2003, http://www.nytimes.com/2003/09/03/education/03CHEA.html.

43. Glynda Hull and Mike Rose, "Rethinking Remediation: Toward a Social-Cognitive of Problematic Reading and Writing," *Written Communication* 6, no 2 (1989): 139–154.

44. Lisa Ede and Andrea Lunsford, *Singular Texts/Plural Authors: Perspectives on Collaborative Writing* (Carbondale: Southern Illinois University Press, 1990).

45. Andrea Lunsford, Rebecca Rickly, Michael J. Salvo, and Susan West, "What Matters Who Writes? What Matters Who Responds? Issues of Ownership in the Writing Classroom," *Kairos: A Journal for Teachers of Writing in Webbed Environments* 1, no. 1 (1996), http://english.ttu.edu/kairos/1.1/features/lunsford/title.html.

46. Linda S. Bergmann, "Higher Education Administration Ownership, Collaboration, and Publication," in *Who Owns This Text? Plagiarism, Authorship, and Disciplin-*

ary Cultures, ed. Carol Peterson Haviland and Joan A Mullin (Logan: Utah State University Press, 2009).

47. Rebecca Moore Howard and Amy E. Robillard, *Pluralizing Plagiarism: Identities, Contexts, Pedagogies* (Portsmouth, NH: Boynton/Cook Publishers, 2008).

48. "2013 Resolutions and Sense of the House Motions," http://www.ncte.org/cccc/resolutions/2013.

49. "NYU Prof Vows Never to Probe Cheating Again—and Faces a Backlash," *Chronicle of Higher Education*, http://chronicle.com/blogs/wiredcampus/nyu-prof-vows-never-to-probe-cheating-again%E2%80%94and-faces-a-backlash/32351.

50. "The Citation Project," http://site.citationproject.net/.

51. Bear Braumoeller and Brian Gaines, "Actions Do Speak Louder than Words: Deterring Plagiarism with the Use of Plagiarism-Detection Software," *Political Science Online*, APSANet (December 2001), http://www.apsanet.org/imgtest/psdec01 braumoellergaines.pdf.

52. Quotations listed on the Center for Academic Integrity website included an admonition to "transcend political correctness," http://www.academicintegrity.org/quotes.asp (accessed September 23, 2005; no longer available online).

53. See Ellen Strenski and Elizabeth Losh, "Initial Report on Plagiarism.org," internal document, University of California, Irvine, 1999.

54. Carol Peterson Haviland and Joan A. Mullin, *Who Owns This Text? Plagiarism, Authorship, and Disciplinary Cultures* (Logan: Utah State University Press, 2009).

55. In our initial experience using Turnitin, we discovered no plagiarizers and announced in our report that we believed it served a "deterrent" function. Subsequently, we found many cases of plagiarism with Turnitin and were persuaded that the early results, based on a limited sample, may have been a fluke.

56. See "Defining and Avoiding Plagiarism: The WPA Statement on Best Practices," Council of Writing Program Administrators, http://www.wpacouncil.org/node/9.

57. David Kay of the School of Information and Computer Science at the University of California, Irvine wrote the initial version of this document for his own upper-division technical communication class.

58. "Welcome to the Virtual Research Home Page," http://e3.uci.edu/faculty/losh/research/.

59. "Virtualpolitik: Obstacles to Collaboration in the Digital University," Research and Occasional Papers Series, Center for Studies in Higher Education (June 1, 2005), Center for Studies in Higher Education. Paper CSHE-9–05, http://repositories.cdlib.org/cshe/CSHE-9-05.

60. Jon Wiener in *Historians in Trouble: Plagiarism, Fraud, and Politics in the Ivory Tower* (New York: New Press, 2005) argues that historians from the left have faced more public shaming and stiffer academic penalties than those who represent conservative interests.

61. Quoted in Grace Lee, "Plagiarism 101," *Read Me: New Media. Net Culture. Now,* http://journalism.nyu.edu/pubzone/ReadMe/article.php?id=441.

62. Andrea L. Foster, "Plagiarism Detection Tool Creates Legal Quandary," *Chronicle of Higher Education,* May 17, 2002, http://chronicle.com/free/v48/i36/36a03701.htm.

63. Stephanie Vie, "Turn It Down, Don't Turnitin: Resisting Plagiarism Detection Services by Talking about Plagiarism Rhetorically," *Computers and Composition Online,* spring 2013, http://www.bgsu.edu/departments/english/cconline/spring2013_special_issue/Vie/history.html.

64. "Paper Authorship Integrity Research," Center for Information Technology and Society, UCSB, http://www.pairwise.cits.ucsb.edu/.

65. "New Lexis-Nexis Copyguard Combats Growing Problem of Unauthorized Use of Copyrighted Material," Lexis-Nexis Media Relations, August 22, 2005 news release, http://www.lexisnexis.com/about/releases/0820.asp.

66. "About iThenticate | Plagiarism Detection Software," http://www.ithenticate.com/about/.

67. "Plagiarism: Conservative or Liberal?" http://www.ithenticate.com/plagiarism-detection-blog/bid/53171/Plagiarism-Conservative-or-Liberal.

68. "Aaron Swartz's Legacy," *HuffPost Live,* http://live.huffingtonpost.com/r/segment/should-academic/50f3382402a7600b9400023c.

69. "Links Scraped from Twitter Hashtag #pdftribute," http://pdftribute.net.

70. Scott Jaschik, "Buying Its Way Onto the Program?," Inside Higher Ed, May 2, 2008, http://www.insidehighered.com/news/2008/05/02/turnitin.

71. Dànielle Nicole DeVoss and James E. Porter, "Why Napster Matters to Writing: Filesharing as a New Ethic of Digital Delivery," *Computers and Composition* 23, no. 2 (2006): 178–210.

72. See websites such as "SHARE RIGHT: Resources about filesharing," http://www.universityofcalifornia.edu/shareright/, for digital ephemera related to the University of California system's carefully orchestrated public-relations campaign.

73. Bill Marsh, *Plagiarism: Alchemy and Remedy in Higher Education* (Albany: SUNY Press, 2007), 38.

74. "Turnitin for Admissions: Welcome," http://www.turnitinadmissions.com/.

75. John G. Palfrey and Urs Gasser, *Born Digital: Understanding the First Generation of Digital Natives* (New York: Basic Books, 2008). As Palfrey and Gasser point out, social media firms aggregate data from many sources without user consent as well.

76. Evan Buswell, Scott Dexter, Craig Dietrich, and Elizabeth Losh, "Week 1: Ethics," *CCS Working Group 2012*, http://wg12.criticalcodestudies.com/discussion/10/week-1-ethics#Item_49.

77. Roberto Esposito, *Communitas: The Origin and Destiny of Community* (Stanford, CA: Stanford University Press, 2010).

78. Heath Raftery, "Anti-anti-plagiarism," *Killing Mind*, July 22, 2004, http://heath.hrsoftworks.net/archives/000027.html.

79. Turnitin, "Usage Policy," http://turnitin.com/en_us/about-us/privacy-center/usage-policy.

80. Noah Wardrip-Fruin, *Expressive Processing* (Cambridge, MA: MIT Press, 2009).

81. "Board of Directors," http://www.iparadigms.com/board.

7 Toy Problems: Education as Product

1. Adrian Mackenzie, *Wirelessness: Radical Empiricism in Network Cultures* (Cambridge, MA: MIT Press, 2010), 69.

2. Ellen Strenski and Stephen D. Franklin, *Building University Electronic Educational Environments: IFIP TC3 WG3.2/3.6 International Working Conference on Building University Electronic Educational Environments, August 4–6, 1999, Irvine, California, USA* (Dordrecht: Kluwer Academic, 2000).

3. Susan DiRenzo, "A Wireless Laptop-Lending Program: The University of Akron Experience," *Technical Services Quarterly* 20, no. 2 (2002): 1–12.

4. William G., Griswold, Patricia Shanahan, Steven W. Brown, Robert Boyer, Matt Ratto, R. Benjamin Shapiro, and Tan Minh Truong, "ActiveCampus: Experiments in Community-Oriented Ubiquitous Computing," *Computer* 37, no. 10 (October 2004): 73–81.

5. Eric Gordon and Adriana de Souza e Silva, *Net Locality: Why Location Matters in a Networked World* (Malden, MA: Wiley-Blackwell, 2011).

6. See Steve Mann, Jason Nolan, and Barry Wellman, "Sousveillance: Inventing and Using Wearable Computing Devices for Data Collection in Surveillance Environments," *Surveillance & Society* 1, no. 3 (January 9, 2002): 331–355, and Jason Nolan, Steve Mann, and Barry Wellman, "Sousveillance: Wearable and Digital Tools in Surveilled Environments," in *Small Tech: The Culture of Digital Tools*, ed. Byron Hawk, David M. Rieder, and Ollie O. Oviedo (Minneapolis: University of Minnesota Press, 2008).

7. "Sousveillance Grid," http://www.specflic.net/modules/sg.html.

8. Matt Ratto, R. Benjamin Shapiro, Tan Minh Truong, and William G. Griswold, "The ActiveClass Project: Experiments in Encouraging Classroom Participation," n.d., http://cseweb.ucsd.edu/~wgg/Abstracts/activeclass-cscl03.pdf, 6.

9. Jason Farman, *Mobile Interface Theory: Embodied Space and Locative Media* (New York: Routledge, 2012).

10. "Adriene Jenik Active Campus," http://www.adrienejenik.net/activecampus.html.

11. William Griswold, Robert Boyer, Steven W. Brown, and Tan Minh Truong, *Using Mobile Technology to Create Opportunistic Interactions on a University Campus*, n.d., http://cseweb.ucsd.edu/~wgg/Abstracts/ubi-spont02.pdf, 3.

12. "Calit2: Bill Griswold's Adventures in Transparent Infrastructure." *Calit2 Newsroom—Web Articles*, October 15, 2002, http://www.calit2.net/newsroom/article.php?id=205.

13. Paul Dourish and Genevieve Bell, *Divining a Digital Future: Mess and Mythology in Ubiquitous Computing* (Cambridge, MA: MIT Press, 2011), 4.

14. Geoffrey C. Bowker and Susan Leigh Star, *Sorting Things Out: Classification and Its Consequences* (Cambridge, MA: MIT Press, 1999), 34.

15. Susan Leigh Star and Karen Ruhleder, "Steps toward an Ecology of Infrastructure: Design and Access for Large Information Spaces," *Information Systems Research: ISR: A Journal of the Institute of Management Sciences* 7, no. 1 (1996): 111–134.

16. Susan Leigh Star, "The Ethnography of Infrastructure," *American Behavioral Scientist* 43, no. 3 (1999): 380.

17. Star and Ruhleder, "Steps toward an Ecology of Infrastructure."

18. Philip E. Agre, "Infrastructure and Institutional Change in the Networked University," *Information, Communication & Society* 3, no. 4 (2000): 500.

19. Ibid., 502–503.

20. Armando Fox, Brad Johanson, Pat Hanrahan, and Terry Winograd, "Integrating Information Appliances into an Interactive Workspace," *IEEE Computer Graphics and Applications* 20, no. 3 (May 2000): 56.

21. Melissa Gregg, *Work's Intimacy* (Cambridge: Polity, 2011).

22. Lev Manovich, *The Language of New Media* (Cambridge, MA: MIT Press, 2002), 96.

23. Griswold, Boyer, Brown, and Truong, *Using Mobile Technology to Create Opportunistic Interactions*, 3.

24. Ibid.

25. Ibid., 1.

26. Sherry Turkle, *Alone Together: Why We Expect More from Technology and Less from Each Other* (New York: Basic Books, 2011).

27. Sherry Turkle, *The Second Self: Computers and the Human Spirit* (New York: Simon & Schuster, 1984).

28. For more recent work that adopts an optimistic approach to blending online and offline social networks, see Harrison Rainie and Barry Wellman, *Networked: The New Social Operating System* (Cambridge, MA: MIT Press, 2012). They also note that, in their own predigital experiences, traditional colleges were much more segregated by religion, race, class, and gender than are contemporary campuses.

29. Alexander R. Galloway, *The Interface Effect* (Cambridge: Polity, 2012), 120–121.

30. Griswold, Boyer, Brown, and Truong, *Using Mobile Technology to Create Opportunistic Interactions*, 3.

31. Ratto et al., "The ActiveClass Project," 8.

32. William G., Griswold, Patricia Shanahan, Steven W. Brown, Robert Boyer, Matt Ratto, R. Benjamin Shapiro, and Tan Minh Truong, "ActiveCampus: Experiments in Community-Oriented Ubiquitous Computing," *Computer* 37, no. 10 (October 2004): 77–78.

33. Tara Zepel, e-mail to the author, July 6, 2013.

34. "Duke University iPod First-Year Experience Final Evaluation Report," 2005, http://cit.duke.edu/pdf/reports/ipod_initiative_04_05.pdf.

35. Travis Kaya, "Classroom iPad Programs Get Mixed Response," *Chronicle of Higher Education, Wired Campus*, September 20, 2010, http://chronicle.com/blogPost/blogPost-content/27046/.

36. Nicholas Bonsack, "Seton Hill to Give iPads to All Full-Time Students." *Macworld*, March 30, 2010, http://www.macworld.com/article/1150165/setonhill_ipad.html.

37. "Mobile Learning at Seton Hill University 2010–2011," http://www-2011.setonhill.edu/techadvantage/MobileLearningBrochure_CombinedFINAL.pdf.

38. K. Walsh, "Seton Hill University's iPad Rollout—More Insights from a Model Implementation," *EmergingEdTech*, August 21, 2011, http://www.emergingedtech.com/2011/08/seton-hill-universitys-ipad-rollout-a-model-implementation/.

39. Dennis Jerz, "Classroom iPad Programs Get Mixed Response," *Chronicle of Higher Education, Wired Campus*, reposted on *Jerz's Literacy Weblog*, September 21, 2010, http://jerz.setonhill.edu/blog/2010/09/21/classroom_ipad_programs_get_mi/.

40. "Honors College University of Maryland," http://www.honors.umd.edu/newsletter201109_dcc.php.

41. Jason Farman, "Encouraging Distraction? Classroom Experiments with Mobile Media," *Chronicle of Higher Education, ProfHacker*, February 9, 2012, http://chronicle.com/blogs/profhacker/encouraging-distraction-classroom-experiments-with-mobile-media/38454.

42. Anne Friedberg, *The Virtual Window: From Alberti to Microsoft* (Cambridge, MA: MIT Press, 2006).

8 The Play's the Thing: Games and Virtual Worlds in Higher Education

1. Tom Boellstorff, *Coming of Age in Second Life: An Anthropologist Explores the Virtually Human* (Princeton, NJ: Princeton University Press, 2008).

2. Pierre Baldi and Crista Lopes, "The Universal Campus: An Open Virtual 3-D World Infrastructure for Research and Education," *eLearn* 2012, no. 4 (April 2012).

3. Elizabeth Losh, "Second Draft Second Life," *Virtualpolitik*, October 22 2007, http://virtualpolitik.blogspot.com/2007/10/second-draft-second-life.html.

4. James A. Inman, *Computers and Writing: The Cyborg Era* (New York: Routledge, 2004).

5. Suzanne C. Baker, Ryan K. Wentz, and Madison M. Woods, "Using Virtual Worlds in Education: Second Life® as an Educational Tool," *Teaching of Psychology* 36, no. 1 (2009): 59–64.

6. Beth Sussman, "Teachers, Colleges Lead a Second Life: Virtual World Can Be a Good Place to Study," *USA Today*, Aug 2, 2007, http://www.usatoday.com/educate/college/arts/articles/20070812.htm.

7. "Vassar's Virtual Sistine Chapel—Vassar College," http://www.vassar.edu/headlines/2007/sistine-chapel.html.

8. "Second Life Tour: Annenberg Island," January 16, 2007, http://www.youtube.com/watch?v=_iMWzuWD28g&.

9. David F. Noble, *Digital Diploma Mills: The Automation of Higher Education* (New York: Monthly Review Press, 2001).

10. Mark Nunes, *Cyberspaces of Everyday Life* (Minneapolis: University of Minnesota Press, 2006), 131.

11. Jessica Hodgins, Amy Bruckman, Paul Hemp, Cory Ondrejka, and Vernor Vinge, "The Potential of End-User Programmable Worlds: Present and Future," in *ACM SIGGRAPH 2007 Panels* (New York: ACM Press, 2007).

12. Amy Bruckman, "Cyberspace Is Not Disneyland: The Role of the Artist in a Networked World," 1995, http://www.cc.gatech.edu/elc/papers/bruckman/disneyland-bruckman.pdf.

13. Shaowen Bardzell, "The Submissive Speaks: The Semiotics of Visuality in Virtual BDSM Fantasy Play," paper presented at Sandbox SIGGRAPH Symposium, July 29, 2006.

14. Interview by the author with Mark Bell.

15. Julian Dibbell, *My Tiny Life: Crime and Passion in a Virtual World* (New York: Holt, 1998).

16. Michael Bugeja, "Avatar Rape," *Inside Higher Ed*, February 25, 2010, http://www.insidehighered.com/views/2010/02/25/bugeja.

17. Clay Shirky, "Second Life: A Story Too Good to Check," Valleywag, December 12, 2006, http://valleywag.com/tech/second-life/a-story-too-good-to-check-221252.php.

18. Henry Jenkins, "A Second Look at Second Life," Confessions of an Aca/Fan Weblog, January 30, 2007, http://www.henryjenkins.org/2007/01/a_second_look_at_second_life.html.

19. Ibid.

20. Brock Read, "Blackboard Video Mocks Second Life," *Chronicle of Higher Education, Wired Campus*, July 6, 2007, http://chronicle.com/blogs/wiredcampus/blackboard-video-mocks-second-life/3164.

21. "Adventures in First Life Redux," June 13, 2007, http://www.youtube.com/watch?v=JprFMs5vKak&.

22. For some of the coverage of the Woodbury events, see Jeffrey R. Young, "Woodbury U. Banned from Second Life, Again," *Chronicle of Higher Education, Wired Campus*, April 21, 2010, http://chronicle.com/blogs/wiredcampus/woodbury-u-banned-from-second-life-again/23352, and Ludlow's coverage on Henry Jenkins blog at http://henryjenkins.org/2010/04/watching_the_watchers_power_an.html and http://henryjenkins.org/2010/04/watching_the_watchers_power_an_1.html.

23. "Anteater Island Closure," http://www.lib.uci.edu/features/news/second-life.html.

24. Elizabeth Deeds Ermarth, "Postmodernity Is Not Postmodernism: Implications for Individuality and Agency," paper presented at Modern Language Association, Philadelphia, December 29, 2006.

25. James Paul Gee, "What Is Literacy?" *Journal of Education* 171, no. 1 (1989): 18.

26. "Audacious Speculations (Fixed Audio)," April 24, 2013, http://www.youtube.com/watch?v=2V7xeLBEIa8&.

27. Susana Tosca and Lisbeth Klastrup, "'Because It Just Looks Cool!'—Fashion as Character Performance: The Case of WoW," *Journal for Virtual Worlds Research* 1, no. 3 (2009).

28. Michele White, *The Body and the Screen Theories of Internet Spectatorship* (Cambridge, MA: MIT Press, 2006).

29. Chris Conway, "Professor Avatar," *Inside Higher Ed*, http://www.insidehighered.com/views/2007/10/16/conway.

30. Janine Fron, Tracy Fullerton, Jacquelyn Ford Morie, and Celia Pearce, "Playing Dress-Up: Costumes, Roleplay and Imagination," University of Modena and Reggio Emilia, 2006, http://egg.lmc.gatech.edu/publications/LudicaDress_Up.pdf.

31. Jenkins, "A Second Look at Second Life."

32. Eva L. Baker, "2007 Presidential Address—The End(s) of Testing," *Educational Researcher* 36, no. 6 (2007): 309–317.

33. "What Are Badges?," P2PU Help, https://www.diigo.com/bookmark/http%3A%2F%2Fp2pu.org%2Fen%2Fpages%2Fassessments-and-badges?tab=people&uname=machinicphylum.

34. "Badges," Open.Michigan Wiki, https://open.umich.edu/wiki/Badges.

35. "About," Open Badges, http://openbadges.org/about/.

36. HASTAC SC 2012 Annual Report with Addendum.

37. Agricultural Sustainability Institute, University of California, Davis, "Fiscal Year 2011–12," http://asi.ucdavis.edu/about/funding/fiscal-year-2011-2012.

38. Kevin Carey, "A Future Full of Badges—Commentary," *Chronicle of Higher Education*, April 8, 2012, http://chronicle.com/article/A-Future-Full-of-Badges/131455/.

39. Jeffrey R. Young, "'Badges' Earned Online Pose Challenge to Traditional College Diplomas," *Chronicle of Higher Education*, January 8, 2012, http://chronicle.com/article/Badges-Earned-Online-Pose/130241/.

40. Alex Reid, "Welcome to Badge World," *Digital Digs*, September 15, 2011, http://alex-reid.net/2011/09/welcome-to-badge-world.html.

41. For a less jaded view of casual gaming, see Jesper Juul, *A Casual Revolution: Reinventing Video Games and Their Players* (Cambridge, MA: MIT Press, 2012).

42. Razvan Rughinis, "Talkative Objects in Need of Interpretation. Re-thinking Digital Badges in Education," *CHI '13 Extended Abstracts on Human Factors in Computing Systems*, 2099–2108 (New York: ACM Press, 2013).

43. Mark R. Lepper, David Greene, and Richard E. Nisbett, "Undermining Children's Intrinsic Interest with Extrinsic Reward: A Test of the 'Overjustification' Hypothesis," *Journal of Personality and Social Psychology* (1973), http://www.eric.ed.gov/ERICWebPortal/detail?accno=EJ085275.

44. Edward Deci, "Effects of Externally Mediated Rewards on Intrinsic Motivation," *Journal of Personality and Social Psychology* 18, no. 1 (1971): 105–115.

45. "We Don't Need No Stinkin' Badges: How to Re-invent Reality without Gamification," GDC Vault, http://www.gdcvault.com/play/1014576/We-Don-t-Need-No.

46. "Gamification: Applying Game Principles to Your Teaching," Magna Publications, http://www.magnapubs.com/catalog/gamification-applying-game-principles-to-your-teaching/.

47. "The Gamification of College Lectures at the University of Michigan," http://www.gamification.co/2013/02/08/the-gamification-of-college-lectures-at-the-university-of-michigan/.

48. "Seminar 1P—Charting the Course for Gamification in Higher Education: Strategies for Success (separate Registration Required)," Educause 2013 Annual Conference, http://www.educause.edu/node/287833.

49. "Professional Certificate in Serious Games and Gamification," Elmhurst College, http://public.elmhurst.edu/admission/school_for_professional_studies/certificate_programs/serious_gaming.

50. Nick Pelling, "The (short) Prehistory of 'Gamification'...," *Funding Startups (& Other Impossibilities)*, http://nanodome.wordpress.com/2011/08/09/the-short-prehistory-of-gamification/.

51. David Crookall, "Serious Games, Debriefing, and Simulation/Gaming as a Discipline," *Simulation & Gaming* 41, no. 6 (2010): 898–920.

52. Byron Reeves and J. Leighton Read, *Total Engagement: Using Games and Virtual Worlds to Change the Way People Work and Businesses Compete* (Boston: Harvard Business Press, 2009).

53. Jane McGonigal, *Reality Is Broken: Why Games Make Us Better and How They Can Change the World* (New York: Penguin Press, 2011).

54. Ian Bogost, "Gamification Is Bullshit," *Atlantic*, August 9, 2011, http://www.theatlantic.com/technology/archive/2011/08/gamification-is-bullshit/243338/.

55. Ian Bogost, *Persuasive Games: The Expressive Power of Videogames* (Cambridge, MA: MIT Press, 2007).

56. James Paul Gee, *What Video Games Have to Teach Us about Learning and Literacy* (New York: Palgrave Macmillan, 2003).

57. Ibid.

58. Bogost, *Persuasive Games*.

59. Ian Bogost, Simon Ferrari, and Bobby Schweizer, *Newsgames: Journalism at Play* (Cambridge, MA: MIT Press, 2010), 10–11.

60. Jane McGonigal, *Reality Is Broken: Why Games Make Us Better and How They Can Change the World* (New York: Penguin Press, 2011), 184.

61. Ibid., 152.

62. Ibid., 151.

63. Ibid., 195.

64. Ian Bogost, "A Portrait of the Artist as a Game Studio," *Atlantic*, March 15, 2012, http://www.theatlantic.com/technology/archive/12/03/a-portrait-of-the-artist-as-a-game-studio/254494/.

65. Ian Bogost, *How to Do Things with Videogames* (Minneapolis: University of Minnesota Press, 2011), 14.

66. Ian Bogost, "Reality Is Alright," January 14, 2011, http://www.bogost.com/blog/reality_is_broken.shtml.

67. Jesper Juul, *The Art of Failure: An Essay on the Pain of Playing Video Games* (Cambridge, MA: MIT Press, 2013).

68. Andrea L. Foster, "Virtual World Modeled on Shakespeare's Works on Hold," *Chronicle of Higher Education, Wired Campus*, October 2, 2007, http://chronicle.com/wiredcampus/article/2425/virtual-world-modeled-on-shakespeares-works-on-hold.

69. Edward Castronova, "Terra Nova: Two Releases: Arden I and Exodus," *Terra Nova*, November 27, 2007, http://terranova.blogs.com/terra_nova/2007/11/two-releases-ar.html.

70. Erica Naone, "Virtual Labor Lost: The Failure of a Highly Anticipated, Multiplayer Game Shows the Limits of Academic Virtual Worlds," *Technology Review*, December 5, 2007, http://www.technologyreview.com/Infotech/19817/?a=f.

71. Mia Consalvo, *Cheating: Gaining Advantage in Videogames* (Cambridge, MA: MIT Press, 2007).

72. Scot Osterweil, "Designing Learning Games That Matter," October 1, 2007, http://www.nercomp.org/data/media/ScotO_10.01.07.pdf.

73. Kurt Squire and Henry Jenkins, "Harnessing the Power of Games in Education," *Insight: The Institute for the Advancement of Emerging Technology in Education* 3, no. 5 (2003) http://website.education.wisc.edu/kdsquire/manuscripts/insight.pdf.

74. Zach Whalen, "'Speare: The Literacy Arcade Game," *Gameology*, February 25, 2006, http://www.canadianshakespeares.ca/speare.

75. Willshakes, "'Speare," *The Dead Authors Blog*, August 27, 2008, http://deadauthors.livejournal.com/5295.html.

76. Squire and Jenkins, "Harnessing the Power of Games in Education."

77. Ibid.

78. "Arden: The World of William Shakespeare," FAQ Version 1.2, June 23, 2006, http://swi.indiana.edu/ardenfaq.pdf (accessed May 19, 2008; no longer available online).

79. Naone, "Virtual Labor Lost."

80. Henry Jenkins, *Fans, Bloggers, and Gamers: Exploring Participatory Culture* (New York: New York University Press, 2006).

81. Marjorie B. Garber, *Vested Interests: Cross-Dressing and Cultural Anxiety* (New York: Routledge, 1992).

82. Raph Koster, *A Theory of Fun for Game Design* (Scottsdale, AZ: Paraglyph Press, 2005).

83. Castronova, "Terra Nova: Two Releases: Arden I and Exodus."

84. Ibid.

85. Ian Bogost, "Videogame Adaptation and Translation," http://www.bogost.com/teaching/videogame_adaptation_and_trans.shtml.

86. "GDC Radio: Game Design Challenge: The Emily Dickinson License," *Gamasutra*, http://gamasutra.com/features/20060509/gamapodcast_01.shtml.

87. "Arden: The World of William Shakespeare," FAQ Version 1.2, June 23, 2006, http://swi.indiana.edu/ardenfaq.pdf (accessed May 19, 2008; no longer available online).

88. Edward Castronova, "Arden Slows Down, Takes Breather," *Terra Nova*, October 2, 2007, http://terranova.blogs.com/terra_nova/2007/10/arden-slows-dow.html.

89. Michael A. Anderegg, *Orson Welles, Shakespeare, and Popular Culture* (New York: Columbia University Press, 1999).

90. Jesper Juul, *Half-Real: Video Games between Real Rules and Fictional Worlds* (Cambridge, MA: MIT Press, 2005).

91. Alexander R. Galloway, *Gaming: Essays on Algorithmic Culture* (Minneapolis: University of Minnesota Press, 2006).

92. Ian Bogost, *Unit Operations: An Approach to Videogame Criticism* (Cambridge, MA: MIT Press, 2006).

93. Tim Harford, "Are Online Gamers Normal Economic Agents?" *Undercover Economist,* May 21, 2008, http://blogs.ft.com/undercover/2008/05/are-online-gamers-normal-economic-agents/.

94. Edward Castronova, *A Test of the Law of Demand in a Virtual World: Exploring the Petri Dish Approach to Social Science,* SSRN Scholarly Paper, Rochester, NY: Social Science Research Network, July 1, 2008, http://papers.ssrn.com/abstract=1173642.

95. "Welcome to the Superstruct Game Archive," http://archive.superstructgame.net/home.

96. Dave Szulborski, *This Is Not a Game: A Guide to Alternate Reality Gaming* (Macungie, PA: New-Fiction Publishing, 2005).

97. Beth Winegarner, "Virginia Tech Massacre," *Backward Messages,* January 30, 2012, http://backwardmessages.wordpress.com/tag/virginia-tech-massacre/.

98. Michael Anderson, "IU's 'Skeleton Chase' Gives Students the Runaround," *ARGNet: Alternate Reality Gaming Network,* January 25, 2009, http://www.argn.com/2009/01/ius_skeleton_chase_gives_students_the_runaround/.

99. Benjamin Stokes, Jeff Watson, Tracy Fullerton, and Simon Wiscombe, "A Reality Game to Cross Disciplines: Fostering Networks and Collaboration," paper presented at DiGRA 2013 Defragging Game Studies, Atlanta, GA.

100. Jean Lave and Etienne Wenger, *Situated Learning: Legitimate Peripheral Participation* (Cambridge: Cambridge University Press, 1991) and Etienne Wenger, *Communities of Practice: Learning, Meaning, and Identity* (Cambridge: Cambridge University Press, 1998).

101. Paul Graham, *Hackers and Painters: Big Ideas from the Computer Age* (Farnham: O'Reilly, 2004).

102. Jeff Watson, "Reality Ends Here," http://remotedevice.net/projects/reality/.

103. Ibid.

104. Ryan Gilmour, "Anatomy of a Game," http://remotedevice.net/wp-content/uploads/2011/09/Anatomy-of-a-Game.pdf.

105. "Game Office | Reality," http://reality.usc.edu/game-office/.

106. "60 Seconds to Safety @SCA," 2011, http://vimeo.com/27302402.

107. Ibid.

108. "The Game 2: A Forbidden Deal," December 2, 2011, http://www.youtube.com/watch?v=EJbZh3Chdtk&.

109. "Just Press Play," https://play.rit.edu/.

110. "RIT Interactive Games & Media," April 17, 2012, http://www.youtube.com/watch?v=0byon8PMa78&.

111. Jeff Watson, "Reality Ends Here: Environmental Game Design and Participatory Spectacle," Ph.D. Dissertation, August 2012, University of Southern California, http://remotedevice.net/docs/Watson_Dissertation_2012.pdf.

112. Monte, "Duke Research Blog: 'Emergence' Wages Peace, Not War, After Droid Revolt," *Duke Research Blog*, October 15, 2009, http://dukeresearch.blogspot.com/2009/10/emergence-wages-peace-not-war-after.html.

113. "Speculation," *Franklin Humanities Institute*, http://www.fhi.duke.edu/blog/speculation.

114. For more, see the work of Nick Gessler at others at Duke University and the What If? Project at https://web.duke.edu/isis/gessler/whatif/.

115. Shirley Brice Heath, "Protean Shapes in Literacy Events: Ever-Shifting Oral and Literate Traditions," in *Literacy: A Critical Sourcebook*, ed. Ellen Cushman, Eugene R. Kintgen, Barry Kroll, and Mike Rose (New York: St. Martin's Press, 2001).

116. "Playing to Win in 1980's Poland," PRI's The World, http://www.theworld.org/2013/07/1980s-poland-game/.

9 Gaining Ground in the Digital University

1. "Naked Water Balloon Fight," 2013, https://www.facebook.com/events/471814222893701/?directed_target_id=173708049458513.

2. Kate Bornstein, *My Gender Workbook: How to Become a Real Man, a Real Woman, the Real You, or Something Else Entirely* (New York: Routledge, 1998).

3. Elizabeth Losh, "Hacktivism and the Humanities: Programming Protest in the Era of the Digital University," in *Debates in the Digital Humanities*, ed. Matthew K. Gold (Minneapolis: University of Minnesota Press, 2013).

4. "Sixth College Name Reveal," https://www.facebook.com/events/173708049458513/.

5. Clark Kerr, *The Uses of the University* (Cambridge, MA: Harvard University Press, 2001).

6. Lawrence Lessig, *Code and Other Laws of Cyberspace* (New York: Basic Books, 1999).

7. Lucille Alice Suchman, *Human–Machine Reconfigurations: Plans and Situated Actions* (Cambridge: Cambridge University Press, 2007), 11.

8. During classroom observations of instructors at my own college, I create a simplified student-seating map, with each student marked as an "X." During the observation, I draw a circle around the "X" each time the student who is represented speaks. This helps me identify how many people speak, how often, what percentage of the class is talking, and if there are gender or racial disparities in participation.

9. Stephanie Simon, "Biosensors to Monitor U.S. Students' Attentiveness," *Reuters*, June 13, 2012, http://www.reuters.com/article/2012/06/13/us-usa-education-gates-idUSBRE85C17Z20120613.

10. Melissa Gregg, "'Getting Things Done'®: Productivity, Self-Management, and the Order of Things," http://academia.edu/2403641/_Getting_Things_Done_Productivity_self-management_and_the_order_of_things.

11. Steve Krause, "E-Learning and Digital Cultures Ends with a Meh #edcmooc," March 6, 2013, http://stevendkrause.com/2013/03/06/e-learning-and-digital-cultures-ends-with-a-meh-edcmooc/.

12. *Reinventing Undergraduate Education: A Blueprint for America's Research Universities*, Boyer Commission on Educating Undergraduates in the Research University, 1998, State University of New York, http://www.sunysb.edu/boyerreport; http://www.eric.ed.gov/ERICWebPortal/detail?accno=ED424840.

13. Association of American Colleges and Universities, "High-Impact Practices," http://www.aacu.org/leap/hip.cfm.

14. Diane Harley, *Use and Users of Digital Resources: A Focus on Undergraduate Education in the Humanities and Social Sciences*, University of California, Berkeley: Center for Studies in Higher Education, April 5, 2006, http://cshe.berkeley.edu/research/digitalresourcestudy/report/.

15. Pierre Lévy, *Collective Intelligence: Mankind's Emerging World in Cyberspace* (New York: Plenum Trade, 1997).

16. Anne Balsamo and Steve Anderson, "A Pedagogy for Original Synners," in *Digital Youth, Innovation, and the Unexpected*, ed. Tara McPherson (Cambridge, MA: MIT Press, 2007), 243.

17. Beatriz da Costa and Kavita Philip, *Tactical Biopolitics Art, Activism, and Technoscience* (Cambridge, MA: MIT Press, 2008).

18. Balsamo and Anderson, "A Pedagogy for Original Synners," 243.

19. The classic example of such prototyping is the mock-up for Alan Kay's Dynabook computer, which was built in cardboard for Kay's public demonstrations. See "Alan Kay's Dynabook—Rare NHK Video," December 21, 2009, http://www.youtube.com/watch?v=r36NNGzNvjo&.

20. Cheryl Ball, e-mail to the TechRhet listserv, November 4, 2012.

21. Garnet Hertz, http://www.conceptlab.com/.

22. Elizabeth Losh, "A Professor with Unconventional Methods, Message," *DML Central*, June 8, 2010, http://dmlcentral.net/blog/liz-losh/professor-unconventional-methods-message.

23. Ernest Morrell, *Critical Literacy and Urban Youth: Pedagogies of Access, Dissent, and Liberation* (New York: Routledge, 2008).

24. Elizabeth Losh, "Going Low-Tech to Teach New Literacies," *DML Central*, May 28, 2012, http://dmlcentral.net/blog/liz-losh/going-low-tech-teach-new-literacies.

25. For more about the arbitrary nature of class-time apportionment of in the academic calendar, see Benjamin Bratton's address to the audience of Mobility Shifts: An International Futures of Learning Summit, in which Bratton defended the idea of a course unfolding over the course of centuries.

26. Nick Montfort, "The Cyborg Campus," http://nickm.com/writing/essays/cyborg_campus.html.

27. Roger Whitson, "Hack-a-Thon: Design a DH/Multimodal Degree Program," *THATCamp Hybrid Pedagogy 2012*, October 20, 2012, http://hybridpedagogy2012.thatcamp.org/10/20/hack-a-thon-design-a-dhmultimodal-degree-program/.

28. "Million Syllabi Hackathon Workshop," DHWI Wiki, http://mith.umd.edu/dhwiwiki/index.php/Million_Syllabi_Hackathon_Workshop.

29. Lilly Irani, "The Anthropologist at the Hackathon: Cultures and Ruptures of Innovation Practice," paper presented at American Anthropological Association Annual Meeting, San Francisco, Nov. 16, 2012.

30. Elizabeth Losh, "How to Use Wikipedia as a Teaching Tool: Adrianne Wadewitz," *DMLcentral*, May 6, 2013, http://dmlcentral.net/blog/liz-losh/how-use-wikipedia-teaching-tool-adrianne-wadewitz.

31. FemTechNet, "Feminist Dialogues in Technology WIKI PARTY!" March 8, 2013, http://femtechnet.blogspot.com/2013/03/feminist-dialogues-in-technology-wiki.html.

32. *Editing on Wikipedia*, 2013, http://vimeo.com/64973792.

33. To be fair to the instructor, the assignment was subsequently dramatically revised. See "Teaching Technology: Remix Video," http://justtv.wordpress.com/2008/04/22/teaching-technology-remix-video/.

34. "Values in Design Workshop—2012," *EVOKE Lab*, http://evoke.ics.uci.edu/values-in-design-workshop-2012-2/.

35. Matthew Fuller and Andrew Goffey, *Evil Media* (Cambridge, MA: MIT Press, 2012).

36. Bill Readings's *The University in Ruins* (Cambridge, MA: Harvard University Press, 1996) remains a classic text about dissensus. For more on the subject, see Jacques Rancière and Steve Corcoran, *Dissensus on Politics and Aesthetics* (New York: Continuum, 2010).

37. In large lecture courses I play music as students enter the room. Students who suggest appropriate songs receive themed mixtapes on CD with the branding of the course logo.

38. Having read thousands of student course evaluations, I can attest to the existence of patterns of bias based on gender and race, but those biases are even more strongly marked on review sites where racist, sexist, and homophobic language is not censured, and "hotness" is rewarded.

39. AnnaLee Saxenian, *Regional Advantage: Culture and Competition in Silicon Valley and Route 128* (Cambridge, MA: Harvard University Press, 1994).

40. For more, see Elizabeth Losh, "Play, Things, Rules, and Information: Hybridizing Learning in the Digital University," *Leonardo Electronic Almanac, DAC09: After Media: Embodiment and Context* 17, no. 2 (2012): 86–102.

41. For more on how the design of Facebook may preclude professor—student relationships, see Ian Bogost, "A Professor's Impressions of Facebook," August 19, 2007, http://www.bogost.com/blog/a_professors_impressions_of_fa.shtml.

42. See "Posts from an English Teacher—Funny Facebook Status Messages and Facebook Fails," n.d., http://failbook.failblog.org/2011/01/30/funny-facebook-fails-posts-from-an-english-teacher/ for an example of potential teacher vulnerability, and "Teacher of the Year—Funny Facebook Status Messages and Facebook Fails," n.d., http://failbook.failblog.org/2011/07/28/funny-facebook-fails-teacher-of-the-year/ for humiliation of a student.

43. For example, the Facebook identity for interacting with a superior-friend might be "John Smith," while the Facebook identity for interacting with a peer-friend might be "Nhoj Htims"; the student may maintain and update two totally separate profiles throughout high school and college.

44. As jobseekers, these students also use these sites for references from employers as well as teachers. The classic work on the importance of social "weak ties" for jobseekers continues to be Mark S. Granovetter, "The Strength of Weak Ties," *American Journal of Sociology* 78, no. 6 (2007): 1360–1380.

45. Lahle Wolfe, "Free Ning No More—As of July 2010 All Ning Site Owners Must Pay," http://womeninbusiness.about.com/b/2010/05/04/free-ning-no-more-as-of-july-2010-all-ning-site-owners-must-pay.htm.

46. See James Paul Gee, *The Anti-Education Era* (New York: Palgrave, 2013), 114–115.

47. See Nicholas Christakis and James Fowler, *Connected The Surprising Power of Our Social Networks and How They Shape Our Lives* (New York: Little, Brown, 2009) for more about the theories of contagion from the social sciences presented in Fowler's course.

48. Yochai Benkler, *The Penguin and the Leviathan: The Triumph of Cooperation Over Self-Interest* (New York: Crown Business, 2011).

Index